Research on Cosmic Rays and Their Impact on Human Activities

Research on Cosmic Rays and Their Impact on Human Activities

Editors

Roberta Sparvoli
Matteo Martucci

MDPI • Basel • Beijing • Wuhan • Barcelona • Belgrade • Manchester • Tokyo • Cluj • Tianjin

Editors
Roberta Sparvoli
University of Rome Tor Vergata
Italy

Matteo Martucci
University of Rome Tor Vergata
Italy

Editorial Office
MDPI
St. Alban-Anlage 66
4052 Basel, Switzerland

This is a reprint of articles from the Special Issue published online in the open access journal *Applied Sciences* (ISSN 2076-3417) (available at: https://www.mdpi.com/journal/applsci/special_issues/Cosmic_Rays_Research).

For citation purposes, cite each article independently as indicated on the article page online and as indicated below:

LastName, A.A.; LastName, B.B.; LastName, C.C. Article Title. *Journal Name* **Year**, *Volume Number*, Page Range.

ISBN 978-3-0365-3857-0 (Hbk)
ISBN 978-3-0365-3858-7 (PDF)

© 2022 by the authors. Articles in this book are Open Access and distributed under the Creative Commons Attribution (CC BY) license, which allows users to download, copy and build upon published articles, as long as the author and publisher are properly credited, which ensures maximum dissemination and a wider impact of our publications.

The book as a whole is distributed by MDPI under the terms and conditions of the Creative Commons license CC BY-NC-ND.

Contents

Preface to "Research on Cosmic Rays and Their Impact on Human Activities" ix

Roberta Sparvoli and Matteo Martucci
Advances in the Research on Cosmic Rays and Their Impact on Human Activities
Reprinted from: *Appl. Sci.* 2021, 12, 3459, doi:10.3390/app12073459 1

Michał Karbowiak, Tadeusz Wibig, David Alvarez Castillo, Dmitriy Beznosko, Alan R. Duffy, Dariusz Góra, Piotr Homola, Marcin Kasztelan and Michał Niedźwiecki
Determination of Zenith Angle Dependence of Incoherent Cosmic Ray Muon Flux Using Smartphones of the CREDO Project
Reprinted from: *Appl. Sci.* 2021, 11, 1185, doi:10.3390/app11031185 5

Zeren Zhima, Yunpeng Hu, Xuhui Shen, Wei Chu, Mirko Piersanti, Alexandra Parmentier, Zhenxia Zhang, Qiao Wang, Jianping Huang, Shufan Zhao, Yanyan Yang, Dehe Yang, Xiaoying Sun, Qiao Tan, Na Zhou and Feng Guo
Storm-Time Features of the Ionospheric ELF/VLF Waves and Energetic Electron Fluxes Revealed by the China Seismo-Electromagnetic Satellite
Reprinted from: *Appl. Sci.* 2021, 11, 2617, doi:10.3390/app11062617 17

Matteo Martucci, Roberta Sparvoli, Simona Bartocci, Roberto Battiston, William Jerome Burger, Donatella Campana, Luca Carfora, Guido Castellini, Livio Conti, Andrea Contin, Cinzia De Donato, Cristian De Santis, Francesco Maria Follega, Roberto Iuppa, Ignazio Lazzizzera, Nadir Marcelli, Giuseppe Masciantonio, Matteo Mergé, Alberto Oliva, Giuseppe Osteria, Francesco Palma, Federico Palmonari, Beatrice Panico, Alexandra Parmentier, Francesco Perfetto, Piergiorgio Picozza, Mirko Piersanti, Michele Pozzato, Ester Ricci, Marco Ricci, Sergio Bruno Ricciarini, Zouleikha Sahnoun, Valentina Scotti, Alessandro Sotgiu, Vincenzo Vitale, Simona Zoffoli and Paolo Zuccon
Trapped Proton Fluxes Estimation Inside the South Atlantic Anomaly Using the NASA AE9/AP9/SPM Radiation Models along the China Seismo-Electromagnetic Satellite Orbit
Reprinted from: *Appl. Sci.* 2021, 11, 3465, doi:10.3390/app11083465 35

Victor Getmanov, Vladislav Chinkin, Roman Sidorov, Alexei Gvishiani, Mikhail Dobrovolsky, Anatoly Soloviev, Anna Dmitrieva, Anna Kovylyaeva, Natalya Osetrova and Igor Yashin
Low-Pass Filtering Method for Poisson Data Time Series
Reprinted from: *Appl. Sci.* 2021, 11, 4524, doi:10.3390/app11104524 47

Francesco Palma, Alessandro Sotgiu, Alexandra Parmentier, Matteo Martucci, Mirko Piersanti, Simona Bartocci, Roberto Battiston, William Jerome Burger, Donatella Campana, Luca Carfora, Guido Castellini, Livio Conti, Andrea Contin, Giulia D'Angelo, Cinzia De Donato, Cristian De Santis, Francesco Maria Follega, Roberto Iuppa, Ignazio Lazzizzera, Nadir Marcelli, Giuseppe Masciantonio, Matteo Mergé, Alberto Oliva, Giuseppe Osteria, Federico Palmonari, Beatrice Panico, Francesco Perfetto, Piergiorgio Picozza, Michele Pozzato, Ester Ricci, Marco Ricci, Sergio Bruno Ricciarini, Zouleikha Sahnoun, Valentina Scotti, Roberta Sparvoli, Vincenzo Vitale, Simona Zoffoli and Paolo Zuccon
The August 2018 Geomagnetic Storm Observed by the High-Energy Particle Detector on Board the CSES-01 Satellite
Reprinted from: *Appl. Sci.* 2021, 11, 5680, doi:10.3390/app11125680 59

Krzysztof Gorzkiewicz, Jerzy W. Mietelski, Zbigniew Ustrnul, Piotr Homola, Renata Kierepko, Ewa Nalichowska and Kamil Brudecki
Investigations of Muon Flux Variations Detected Using Veto Detectors of the Digital Gamma-rays Spectrometer
Reprinted from: *Appl. Sci.* **2021**, *11*, 7916, doi:10.3390/app11177916 75

Ernesto Ortiz, Blanca Mendoza, Carlos Gay, Victor Manuel Mendoza, Marni Pazos and Rene Garduño
Simulation and Evaluation of the Radiation Dose Deposited in Human Tissues by Atmospheric Neutrons
Reprinted from: *Appl. Sci.* **2021**, *11*, 8338, doi:10.3390/app11188338 85

Igor Lebedev, Anastasia Fedosimova, Andrey Mayorov, Pavel Krassovitskiy, Elena Dmitriyeva, Sayora Ibraimova and Ekaterina Bondar
Direct Measurements of Cosmic Rays (TeV and beyond) Using an Ultrathin Calorimeter: Lessening Fluctuation Method
Reprinted from: *Appl. Sci.* **2021**, *11*, 11189, doi:10.3390/app112311189 93

Giuseppe Di Sciascio
Measurement of Energy Spectrum and Elemental Composition of PeV Cosmic Rays: Open Problems and Prospects
Reprinted from: *Appl. Sci.* **2022**, *12*, 705, doi:10.3390/app12020705 103

Preface to "Research on Cosmic Rays and Their Impact on Human Activities"

Radiation that comes from outside of the Earth continuously interacts with us, even if we do not even notice it. More often than we think, its presence can affect many aspects of human activities. Some of the latest studies on cosmic rays, both those that are low and high energy, are described in this book, providing a snapshot of the latest research in the scientific field and beyond.

Roberta Sparvoli and Matteo Martucci
Editors

Editorial

Advances in the Research on Cosmic Rays and Their Impact on Human Activities

Roberta Sparvoli [1,2,*] and Matteo Martucci [1,2]

1. Department of Physics, University of Rome "Tor Vergata", V. della Ricerca Scientifica 1, I-00133 Rome, Italy; matteo.martucci@roma2.infn.it
2. INFN—Sezione di Roma "Tor Vergata", V. della Ricerca Scientifica 1, I-00133 Rome, Italy
* Correspondence: roberta.sparvoli@roma2.infn.it

Citation: Sparvoli, B.; Martucci, M. Advances in the Research on Cosmic Rays and Their Impact on Human Activities. *Appl. Sci.* **2021**, *12*, 3459. https://doi.org/10.3390/app1207 3459

Received: 21 March 2022
Accepted: 23 March 2022
Published: 29 March 2022

Publisher's Note: MDPI stays neutral with regard to jurisdictional claims in published maps and institutional affiliations.

Copyright: © 2021 by the authors. Licensee MDPI, Basel, Switzerland. This article is an open access article distributed under the terms and conditions of the Creative Commons Attribution (CC BY) license (https:// creativecommons.org/licenses/by/ 4.0/).

1. Introduction

It is well known that the galactic cosmic-ray spectrum extends over 14 orders of magnitudes in energy and about 12 in intensity, and the detection methods can be divided into two classes. First, there is the "direct detection" of the primary cosmic rays in space or at high altitude, which includes experiments on stratospheric balloons, satellites, etc. Second, there is the "indirect detection" of secondary particles, namely, the extensive air showers produced by a primary cosmic-ray particle impinging the atmosphere. The first method is more adapted to studying the low-energy portion of the spectrum, while the second one is more suited to investigating the region at high or even ultra-high energies. Moreover, while low-energy particles have been more easily studied in the past, their variability in time (mostly linked to solar activity) is continuously challenging scientists, who are trying to model such variation to asses potential risks for human health and activities on the ground and in space. On the other hand, high-energy particles (linked to a galactic or extra-galactic origin) are more difficult to measure, due to the large sensitive areas required to obtain some statistical significance, but they are somewhat more dangerous and show a lower degree of variation.

During the last few decades, new experiments with advanced techniques have been looking to unveiling the properties of cosmic radiation, both at low and high energies. In this Special Issue, both direct and indirect measurements are presented, coming from experiments in data collection or already completed data collection. Emphasis is placed on low-energy electrons and protons detected in flight, and during geomagnetic storms. As for indirect detection, the muon flux determination and modulation at ground are studied in great detail. Some of the most interesting results are presented, and a couple of new techniques in cosmic-ray detection reported.

2. Outlook

The Special Issue starts with several articles dealing with direct measurements of cosmic rays in flight, in different periods of the solar activity.

The variability of the low-energy particle populations (mostly electrons and protons in the sub-Mev or MeV energy range) during the strong geomagnetic storm of 25–26 August 2018 is discussed in [1,2].

In [1], the temporal and spatial distributions of the extremely/very low frequency (ELF/VLF) wave activities and the energetic electron fluxes in the ionosphere are described. This work is based on the observations by a set of detectors onboard the China Seismo-Electromagnetic Satellite (CSES-01). It is shown that the energetic electrons at energies below 1.5 MeV get strong enhancements during the whole storm time on both the day and night side. Moreover, a good correlation of the ionospheric ELF/VLF wave activities with energetic electron precipitations during the various storm evolution phases is revealed.

Variations in the precipitating fluxes are also spotted in correspondence with changing geomagnetic activity.

In [2], the electron rates from the High-Energy Particle Detector (HEPD-01), one of the main payloads onboard the CSES-01 satellite, are studied during the whole period of the same August storm. It has been found that the rate of electrons in the MeV energy range is characterized by a depletion during the storm's main phase and by a clear enhancement during the recovery, caused by large sub-storm activity, with the key role played by auroral processes mapped into the outer belt. A post-storm rate increase is localized at L-shells immediately above ∼3, mostly driven by non-adiabatic local acceleration—caused by possible resonant interaction with low-frequency magnetospheric waves.

In the work by Martucci M. et al. [3], the inner radiation environment—better known as South Atlantic Anomaly—is studied comparing data from the aforementioned High-Energy Particle Detector with the NASA AP9 radiation model at Low-Earth Orbit. This model provides useful information on the energetic protons in the near-Earth environment, but it is still largely incomplete as to some features. The estimation reported in this analysis will serve as the starting ground for a forthcoming extensive testing and validation of other current theoretical and empirical models.

Remaining in the field of direct detection, a new method that makes it possible to use an ultra-thin calorimeter for direct measurements of cosmic rays with energies of TeV and higher is shown in [4]. Due to large fluctuations in shower development, the low statistics of the analyzed events and the large size required for the calorimeter, make it almost impossible to determine the primary energy of an incoming particle. A solution to these problems is proposed on the basis of a lessening fluctuation technique, based on the assumption of the universality of the development of cascades initiated by particles of the same energy and mass. The size of the cascade and the rate of its development are analyzed and the whole method was tested using the calorimeter of the PAMELA collaboration, showing that the primary energy can be determined on the ascending branch of the cascade curve, solving the problems associated with the need to increase the thickness of the detector and with the limitation of the analyzed events.

Touching then the topic of the impact of the radiation on the human activities, the study of the dose absorbed from more heavy particles by spacecraft and crews in a certain radiation environment is crucial to understand the real risks linked to space-flights. For example, in [5], the radiation dose deposited by atmospheric neutrons in human tissues is evaluated. The goal of this work is to obtain the overall dose that atmospheric neutrons (with energy from 1 to 1000 MeV) deposit in tissues of the human body, which means blood, adipose, bone and brain, as a function of both altitude and latitude. With the help of the Geant 4 software, a numerical simulation is developed. The analysis of the atmospheric neutron fluxes obtained from the Excel-Based Program for Calculating Atmospheric Cosmic-Ray Spectrum (EXPACS) shows that the dose deposited by these neutral particles increases with the increase in altitude and latitude, e.g., for an altitude of high mountain (4 km), the dose is increased ∼19 times; while, for an altitude of commercial flights (10 km), it is increased ∼156 times.

For what concerns the higher energies, the increased dangers related to the augmented penetration power of particles is balanced by the relatively low fluxes. Nevertheless, the nature of these particles is not fully understood, leaving many questions unanswered.

In the work reported in [6] by Di Sciascio G., the detection of galactic cosmic rays from ground with air shower arrays up to 10^{18} eV is described. The aim of this paper is to discuss the conflicting results in the 10^{15} eV energy range and the perspectives to clarify the origin of the so-called 'knee' in the all-particle energy spectrum, crucial to give a solid basis for models up to the end of the cosmic ray spectrum. The basic techniques used in reconstructing primary particle characteristics (energy, mass, and arrival direction) from the ground are provided, highlighting why indirect measurements are difficult and results are still conflicting.

Entering the topic of the characterization of the indirect detection on ground, three papers propose different techniques for the muon flux analysis. Some new insights on cosmic-ray muons are reported in [7]. In this paper, the authors present some interesting results on these particles registered by a digital gamma-ray spectrometer's active shield, made of five large plastic scintillators. In analogous active shields working in anticoincidence mode with germanium detectors, the generated data are used only as a gating signal and are consequently not stored. However, thanks to digital acquisition applied in designed novel gamma-ray spectrometers, it has become possible to use generated data to reduce the germanium detector background (cosmic rays veto system) and also to initialize long-term monitoring of the muon flux intensity. Fourier analyses also reveal the presence of daily (24 h), near-monthly (27 days) and over bi-monthly (68 days) cycles.

Problems of digital processing of Poisson-distributed data time series from various counters of radiation particles, photons, slow neutrons etc. are relevant for experimental physics and measuring technology. In [8] a low-pass filtering method for normalized Poisson-distributed data time series is proposed and a digital quasi-Gaussian filter with a finite impulse response is designed. Moreover, the results of testing such filtering method on model and experimental Poisson data from the URAGAN muon hodoscope, are presented.

Finally, the Cosmic-Ray Extremely Distributed Observatory (CREDO) project—established to detect ultra high-energy cosmic ray particles—is described in [9]. Among other and more conventional detection techniques, it makes also use of cameras in smartphones as particle detectors, creating a extremely innovative and highly educational method of experimental observations. In this paper the search for cosmic-ray muons, recorded using this method, is presented.

Funding: This research received no external funding.

Conflicts of Interest: The authors declare no conflict of interest.

References

1. Zhima, Z.; Hu, Y.; Shen, X.; Chu, W.; Piersanti, M.; Parmentier, A.; Zhang, Z.; Wang, Q.; Huang, J.; Zhao, S.; et al. Storm-Time Features of the Ionospheric ELF/VLF Waves and Energetic Electron Fluxes Revealed by the China Seismo-Electromagnetic Satellite. *Appl. Sci.* **2021**, *11*, 2617. [CrossRef]
2. Palma, F.; Sotgiu, A.; Parmentier, A.; Martucci, M.; Piersanti, M.; Bartocci, S.; Battiston, R.; Burger, W.J.; Campana, D.; Carfora, L.; et al. The August 2018 Geomagnetic Storm Observed by the High-Energy Particle Detector on Board the CSES-01 Satellite. *Appl. Sci.* **2021**, *11*, 5680. [CrossRef]
3. Martucci, M.; Sparvoli, R.; Bartocci, S.; Battiston, R.; Burger, W.J.; Campana, D.; Carfora, L.; Castellini, G.; Conti, L.; Contin, A.; et al. Trapped Proton Fluxes Estimation Inside the South Atlantic Anomaly Using the NASA AE9/AP9/SPM Radiation Models along the China Seismo-Electromagnetic Satellite Orbit. *Appl. Sci.* **2021**, *11*, 3465. [CrossRef]
4. Lebedev, I.; Fedosimova, A.; Mayorov, A.; Krassovitskiy, P.; Dmitriyeva, E.; Ibraimova, S.; Bondar, E. Direct Measurements of Cosmic Rays (TeV and beyond) Using an Ultrathin Calorimeter: Lessening Fluctuation Method. *Appl. Sci.* **2021**, *11*, 11189. [CrossRef]
5. Ortiz, E.; Mendoza, B.; Gay, C.; Mendoza, V.M.; Pazos, M.; Garduno, R. Simulation and Evaluation of the Radiation Dose Deposited in Human Tissues by Atmospheric Neutrons. *Appl. Sci.* **2021**, *11*, 8338. [CrossRef]
6. Di Sciascio, G. Measurement of Energy Spectrum and Elemental Composition of PeV Cosmic Rays: Open Problems and Prospects. *Appl. Sci.* **2022**, *12*, 705. [CrossRef]
7. Gorzkiewicz, K.; Mietelski, J.W.; Ustrnul, Z.; Homola, P.; Kierepko, R.; Nalichowska, E.; Brudecki, K. Investigations of Muon Flux Variations Detected Using Veto Detectors of the Digital Gamma-rays Spectrometer. *Appl. Sci.* **2021**, *11*, 7916. [CrossRef]
8. Getmanov, V.; Chinkin, V.; Sidorov, R.; Gvishiani, A.; Dobrovolsky, M.; Soloviev, A.; Dmitrieva, A.; Kovylyaeva, A.; Osetrova, N.; Yashin, I. Low-Pass Filtering Method for Poisson Data Time Series. *Appl. Sci.* **2021**, *11*, 4524. [CrossRef]
9. Karbowiak, M.; Wibig, T.; Alvarez Castillo, D.; Beznosko, D.; Duffy, A.R.; Góra, D.; Homola, P.; Kasztelan, M.; Niedewiecki, M. Determination of Zenith Angle Dependence of Incoherent Cosmic Ray Muon Flux Using Smartphones of the CREDO Project. *Appl. Sci.* **2021**, *11*, 1185. [CrossRef]

Article

Determination of Zenith Angle Dependence of Incoherent Cosmic Ray Muon Flux Using Smartphones of the CREDO Project

Michał Karbowiak [1], Tadeusz Wibig [1,*], David Alvarez Castillo [2], Dmitriy Beznosko [3], Alan R. Duffy [4], Dariusz Góra [5], Piotr Homola [5], Marcin Kasztelan [6] and Michał Niedźwiecki [7]

1. Faculty of Physics and Applied Informatics, University of Lodz, Pomorska 149/153, 90-236 Łódź, Poland; michal.karbowiak@fis.uni.lodz.pl
2. Bogoliubov Laboratory for Theoretical Physics, Joint Institute for Nuclear Research, Joliot-Curie Str. 6, 141980 Dubna, Russia; alvarez@theor.jinr.ru
3. Department of Chemistry and Physics, Clayton State University, 2000 Clayton State Blvd, Morrow, GA 30260, USA; dmitriybeznosko@clayton.edu
4. Centre for Astrophysics and Supercomputing, Swinburne University of Technology, Hawthorn, VIC 3122, Australia; aduffy@swin.edu.au
5. The Henryk Niewodniczański Institute of Nuclear Physics, Polish Academy of Sciences, Radzikowskiego 152, 31-342 Kraków, Poland; Dariusz.Gora@ifj.edu.pl (D.G.); Piotr.Homola@ifj.edu.pl (P.H.)
6. Astrophysics Division, National Centre for Nuclear Research, 28 Pułku Strzelców Kaniowskich 69, 90-558 Łódź, Poland; mk@zpk.u.lodz.pl
7. Department of Information and Communications Technology, Faculty of Computer Science and Telecommunications, Cracow University of Technology, Warszawska 24st, 31-155 Cracow, Poland; nkg@pk.edu.pl
* Correspondence: t.wibig@gmail.com

Citation: Karbowiak, M.; Wibig, T.; Alvarez Castillo, D.; Beznosko, D.; Duffy, A.R.; Góra, D.; Homola, P.; Kasztelan, M.; Niedźwiecki, M. Determination of Zenith Angle Dependence of Incoherent Cosmic Ray Muon Flux Using Smartphones of the CREDO Project. *Appl. Sci.* **2021**, *11*, 1185. https://doi.org/10.3390/app11031185

Academic Editor: Roberta Sparvoli
Received: 5 January 2021
Accepted: 15 January 2021
Published: 28 January 2021

Publisher's Note: MDPI stays neutral with regard to jurisdictional clai-ms in published maps and institutio-nal affiliations.

Copyright: © 2021 by the authors. Licensee MDPI, Basel, Switzerland. This article is an open access article distributed under the terms and conditions of the Creative Commons Attribution (CC BY) license (https://creativecommons.org/licenses/by/4.0/).

Abstract: The Cosmic-Ray Extremely Distributed Observatory (CREDO) was established to detect and study ultra high-energy cosmic ray particles. In addition to making use of traditional methods for finding rare and extended cosmic ray events such as professional-grade Extensive Air Shower (EAS) arrays, as well as educational 'class-room' detectors, CREDO also makes use of cameras in smartphones as particle detectors. Beyond the primary scientific goal of the CREDO project, to detect Cosmic Ray Ensembles, is the equally important educational goal of the project. To use smartphones for EAS detection, it is necessary to demonstrate that they are capable of effectively registering relativistic charged particles. In this article, we show that the events recorded in the CREDO project database are indeed tracing incoherent cosmic ray muons. The specific observed distribution of zenith angle of charged particle direction corresponds to that expected for muons. It is difficult, if not impossible, to imagine different mechanisms leading to such a distribution, and we believe it clearly demonstrates the suitability of smartphone-based detectors in supporting the more traditional cosmic ray detectors.

Keywords: cosmic rays; Extensive Air Showers; particle detectors; Cosmic Ray Ensembles

1. Introduction

The cosmic ray energy spectrum extends from below ∼100 MeV [1] up to ∼10^{20} eV [2]. The spectrum and composition of cosmic rays for energies up to the "knee" is compatible with diffusive shock acceleration mechanisms [3]. The maximum energy achievable in this shock acceleration process is at about PeV region [4,5]. The types of objects in the Universe that are able to accelerate particles at even higher energies are limited in number. In the case of ultra high-energy cosmic rays, it seems that there are few places capable of accelerating particles above 10^{20} eV: large scale shocks surrounding galaxy clusters, internal or external shocks of starburst-superwinds, Active Galactic Nuclei (AGN) or Gamma-Ray Bursts, AGN flares, jets, magnetars, lobes of giant radio galaxies, see [6] for a review of these. Some of the exotic models can be verified by searching for new, rather unexpected behavior of cosmic rays at the highest energies. One of them is a focus of the Cosmic-Ray Extremely Distributed Observatory (CREDO) Collaboration. This global approach allows the testing of hypothesized events of ultra-high energy cosmic ray 'bunches' observed as simultaneous Extensive Air Showers (EAS) over the entire exposed surface of the Earth: so-called Cosmic Ray Ensembles (CRE) [7–9]. Such a phenomenon has never been seen, but there are several models under which such an event is a possibility.

To observe such events, a system operating on that global scale is required. Due to the extreme geographical scale of the data acquisition, the CREDO Collaboration makes use of non-expert science enthusiasts. They participate in research with their own mobile devices equipped with the CREDO Detector application [10] which enables detection of ionizing radiation using the CMOS sensors.

(with their own smartphones). The idea of using an array of smartphones as cosmic ray/muon detectors is a quite recent possibility—as the density of smartphones per squared km is a significant feasibility parameter [11]. This was realized, to some extent, by the DECO [12,13] and CRAYFIS [14] projects, and even in muon flux determination measurements [15], but at the same time the quality of individual data as well as the idea itself was criticized [16,17].

The possibility of registering an EAS by detectors of such small active area requires a large number of smartphones concentrated in a relatively limited footprint. The typical EAS array of contemporary cosmic ray experiments consists of hundreds of detectors exceeding several square meters in size (e.g., KASCADE in Karlsruhe [18]; MAS at Mount Chacaltaya [19], or GRAPES at Ooty [20]).

To effectively study cosmic ray physics at energies of 10^{19} (10^{20}) eV, the density of smartphones (constantly working taking "dark" frames) should be 5000 (1000) per km^2 [17] or as few as 400 [11], which remains a challenging scale of detectors. To have an EAS smartphone array comparable to the large experiments like Pierre Auger Observatory or Telescope Array, we would require millions of phone cameras across an area of thousands of square kilometers, all permanently 'on'. The goal of registering every EAS from each CRE at close to 100% effectiveness, and measuring its characteristics (total number of particles, their distribution with respect to shower core, and incoming direction) is simply unfeasible. However, it may be possible to detect some signal from any CRE using a coincidence across very distant sets of detectors (smartphones).

Another noted challenge is the contamination of smartphone camera signals by sources other than cosmic ray. In principle, this includes the housing of the camera itself, as a smartphone is not made to be "low radiation background". It typically contains many radioactive nuclei of isotopic composition similar to the Earth's crust, which constantly decay to give signals in the semiconductor camera matrix elements. The U, Th, and K nuclei present, for example, in the walls are obvious sources of the unwanted background [16]. Ultimately we need to determine experimentally whether smartphones can be used as cosmic ray particle detectors, or if this background noise removes the possibility of measurement in practice.

To address this issue, we approach it from the opposite case. We make the assumption that particles, other than cosmic muons, do not have a characteristic distribution; unlike the very well known distribution demonstrated by single cosmic ray muons. In Section 3, we will show that long traces fit perfectly to the known muon relation, so we conclude that other sources of charged particle contamination in this particular region is small, but on the other hand, for short traces it may be significant and cannot be eliminated in practice.

2. Registration and Analysis of the Smartphone Recordings

The cosmic ray CREDO Detector application for smartphones is available freely at Google Play. The active users send their records to the CREDO database. They have access in turn to all the data from other users to download, study, and analyze for their own Citizen Science efforts. In the CREDO database, there are approximately a million images registered with the smartphone camera's lens obscured.

Such recorded images should consist mostly of black pixels. When this condition is met, the CREDO Detector application starts working. If a particle of secondary cosmic radiation, muon (or possibly a particle of local radiation source), passes through the active layer of the smartphone camera it will stimulate some of the pixels. A few to several dozen pixels, distributed in a cluster of shapes that range from circular to extended lines, should then appear brighter on the roughly homogeneous black background. In one 24-h period there can be anything from one to several hundred detections.

Events in which very long traces are visible can potentially be the tracks of cosmic ray muons that passed through the camera at large angles. This possibility is, in theory, easily verifiable. The zenith distribution of such incident angles of single incoherent muons is well established, and is actively measured by small, even portable, muon telescopes often created and operated by students [15,21–24] as a demonstration of their detectors. In this paper, we will show that the majority of CREDO registered events are due to real muons with a recovered zenith angle distribution as expected.

The data used to obtain the results presented below consist of more than 100,000 CREDO database registrations obtained by one (very active) user with only one camera. The CREDO application draws a lot of power, and practically can only be used for extended periods when connected to a charger. Therefore, the cameras actively connected to the CREDO network usually rest permanently in the same place, which means that once placed horizontally they remain in this position permanently, and we do not have to be concerned of blurring the distributions by uncontrolled changes in the camera matrix plane. The registrations are available as PNG files cropped from the whole camera frame to 60×60 pixels around the brightest pixel in the frame.

Some events in the database are caused by the single pixel noise signal near the edge of the camera, and the image does not consist of 60×60 pixels around the brightest pixel. This reduces the number of pictures used in the present analysis from 1×10^5 to 6×10^4. Examples are shown in the top row in Figure 1. Signals are roughly proportional to the ionization energy loss in the particular matrix pixels, but it should be noted that some corrections (unknown in practice) have been applied to the photos by the internal camera software. The box 2D-histograms are given in the middle row in Figure 1 which scale in size with the ionization energy. The bottom row is the process by which the main axis of the "muon tracks" are identified.

2.1. Determination of Noise

For each registered event, we determined the average dark pixel brightness and its standard deviation σ using only regions far from bright pixels (these are either intrinsic 'hot spots' on the chip or from the suspected signal itself). This procedure was then repeated after removing the pixels which exceeded 2σ above the initially-estimated average. At this stage, all pixels containing (potentially) everything associated with the registration of a cosmic ray muons (and 'hot spots') are omitted. Statistics from the, over 3000, remaining pixels in each frame allowed us to obtain a well-defined average brightness of dark pixels and its intrinsic variance. These are considered reliable values attributable to the noise in each recorded event.

2.2. The Elongation Axis

The next step in the analysis was to find the main symmetry axis of the track. First we selected all pixels that exceeded the threshold (equal to the previously estimated average signal of the dark pixel noise, plus $10\times$ its finally determined standard deviation). We have confirmed that using a factor of 5 instead of 10 does not significantly influence the results, as the signal we observe is so much brighter than the average noise value.

For the selected pixels (i.e., those above the dark pixel threshold) we determined the 'main axis' of the track. There are, in principle, many ways to determine this, and we have tested several: inertia ellipses, the Hough algorithm line, the smallest sum of squared distances weighted by squares of the brightness, and even the brightness only. They all give very similar results as the images of the tracks in each picture are clear, and regardless of how they are linearized the 'main axis' of the track is (almost) always the same. The lines are shown in the bottom row in Figure 1.

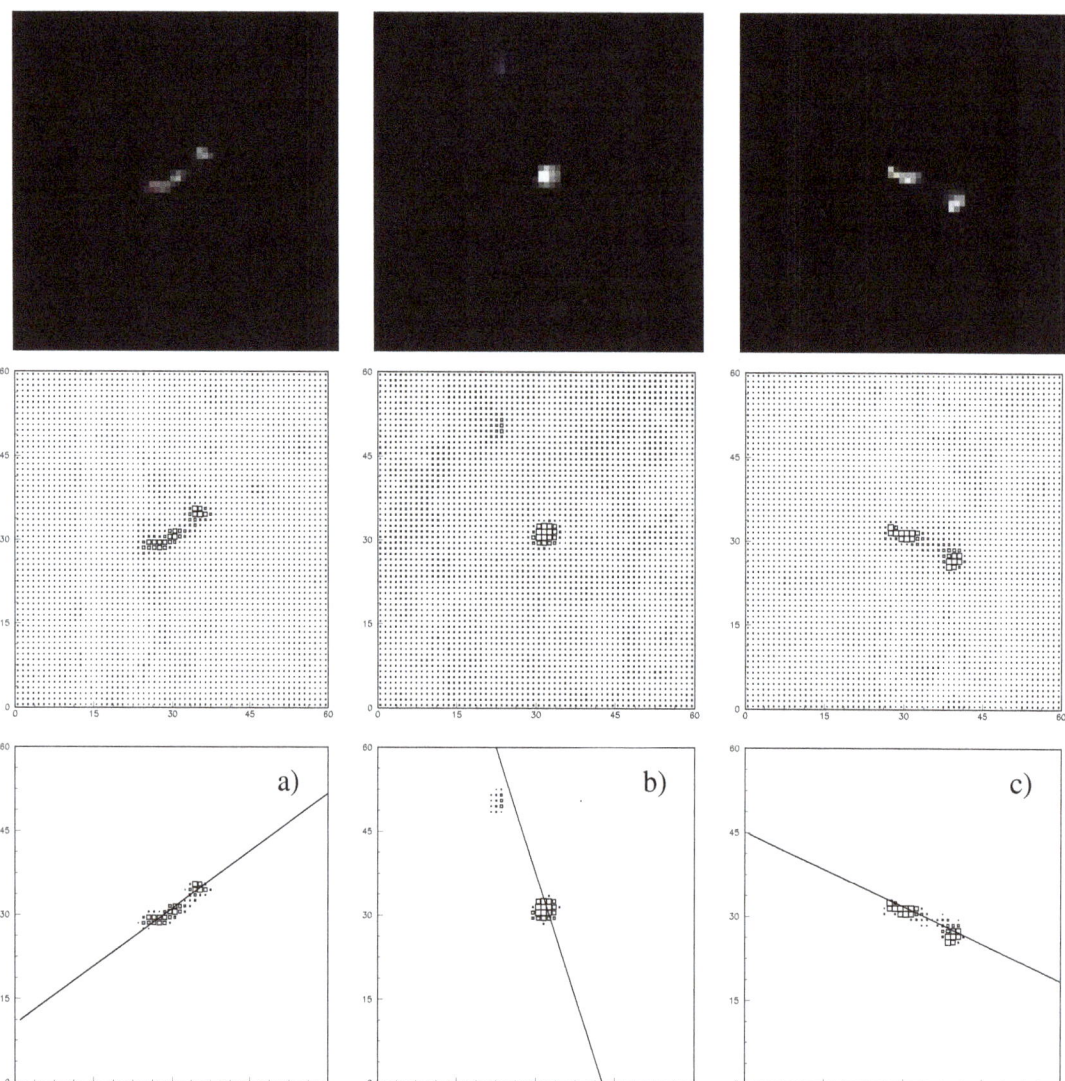

Figure 1. Three example pictures (**a**–**c**) of traces from the CREDO database. The top row shows the original smartphones images, the middle row shows the same events with the size of boxes proportional to the pixel signal (registered light). The bottom row presents the estimated main axis of the "muon tracks".

2.3. The Length of the Track

All pixels exceeding the threshold described in the section before are projected onto the newly-identified main axis of the analyzed trace. Assuming that the axis is the real main axis of the track the obtained histograms represent the ionization along the trace. The respective examples are shown in Figure 2. If we deal with a trace of a relativistic cosmic ray particle (muon), the signal pixels should have roughly the same brightness (with obvious geometric corrections). Such idealized situations are encountered only rarely, however.

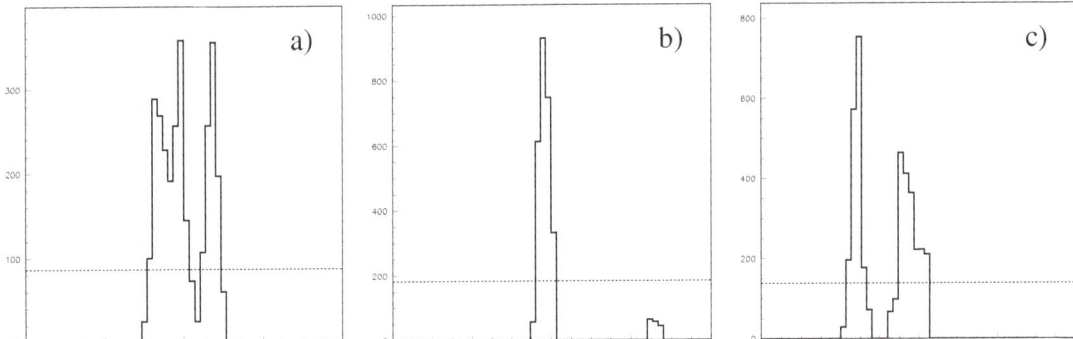

Figure 2. Projections of the pixel signal along the main track axis for the events shown in Figure 1 on (**a**–**c**) plots, respectively. Abscissa is the position along the track axis (in px) and units on the ordinate shows the sum of brightness of the projected pixels (from 0 to 255 for empty and saturated pixel, respectively). The horizontal dashed lines are the values of the cuts used to determine the length of the track (see text for details).

The ends of the track were found by tracking to the left (and to the right) from the point with the highest projected brightness, until the signal values become smaller than limit chosen by trial and error. We defined this limit as a fraction of the second projected brightest bar in the histograms. The values are shown in Figure 2 as dashed lines. Using just the brightest histogram bar value the procedure could be subject to fluctuations to the larger extent. The value of this fraction was selected comparing the results of the algorithm with the subjective perception of the observer. Eventually, the value of 30% was established.

In some cases there is a very small gap in the trace, as it seen in the first (top left) image in the example shown in Figures 1 and 2a, where we can guess that the real track continues on both sides of the two histogram bins with the summed brightness value below the threshold line (Figure 2). Other times, the gap can be quite long as the third column exemplifies, where it extends over 5 histogram bins (Figure 2c). The question then arises if we are dealing with a single track, or a random contamination of unknown origin. Before we attempt to answer this question, we first have to define some relevant variables.

2.4. Zenith Angle of the Particle Track

The zenith angle of the particle track Θ, if we assume that it is indeed the trace left by a particle passing through the entire photosensitive layer of the camera, is naturally determined by the track length observed in the camera plane and thickness of the photosensitive layer of the camera matrix itself.

$$\Theta = \arctan \frac{l}{h} \quad (1)$$

where l is the track length, and h is the depth of the sensitive layer of the matrix in the smartphone camera. Measuring the track length distribution, we can obtain the zenith angle dependence of the observed particle flux. It can then be compared with the prediction of a model. For example, if we assume that particle flux reaches the Earth surface isotropically, the zenith angle distribution observed by the flat horizontally placed detector will be of the form $dN/d(\cos(\Theta)) \sim \cos(\Theta)$. With non-isotropic, but still a power-law in the variable $\cos(\Theta)$, the simple power-law form of the observed distribution is expected.

The cosmic ray single muon zenith inclination angle distribution has been known for nearly 80 years [25,26] and many different measurements confirm that it is quite accurately described as

$$f(\Theta) = \cos^\gamma \Theta \qquad (2)$$

with the index of $\gamma \sim 2$. We will use this simple relation to test our algorithms for observed track identification.

2.5. The Track Length Determination Algorithms

There is, in general, high subjectivity to the interpretation of the images of long tracks with complex and complicated traces. Thankfully, these cases are rare, but they affect the tails of the track length distribution. The algorithm for the track length determination is especially sensitive to the treatment of the multi-component pictures, when the gaps in the traces are seen.

After the projection of the pixel brightness along the identified track's main axis, the gaps in the histogram are clearly visible (see Figure 2) and the question then arises where the real particle track ends, if it is, of course, the real particle track in the first place. This question can be further expanded into more detailed ones: how long can the gaps be allowed in a single track, and how to treat the cases with the longer gaps. In principle, we can accept just the longer length regardless, or we can reject the entire frame from the track length distribution estimation procedure. We have tested several of these possibilities, noting as before that, with the appropriate method, the true cosmic ray muon tracks are expected to follow the power-law distribution of $\cos(\Theta)$.

The results of the six procedures are shown in Figure 3. The respective $\cos(\Theta)$ distributions are shown there. To convert from l to Θ (or to $\cos(\Theta)$), the depth of the camera's photosensitive layer is set here, for the time being, to 5 px. For simplicity we will, hereafter, measure all distances, and the track length, as well as the thickness of the matrix sensitive layer, using as a unit a single matrix pixel size (px). This is quite a natural unit if we are dealing with the same camera, as is in the case of the present work. The thickness of the photosensitive layer will be considered later in this paper.

We explore in detail below the tested algorithms using the examples in Figure 3. Initially, we explore the steps within the least restrictive algorithm:

(a) acceptable gaps in the histogram tracks are of length up to 4 px (as shown in Figure 2) but exactly one track is seen in the picture,
(b) as above up to 4 px gaps are accepted but 'multiple track' events are included.

In the middle row of Figure 3, we compared results of algorithms for 'multiple track' cases

(c) allowing events with large gap (up to 5 px) in the track,
(d) only the small gap (only below 3 px) is acceptable.

In the lower row of Figure 3, we present results for single-track cases and a very strong criterion for the gap length in the track

(e) only 1 px long gaps, at most, are accepted,
(f) no gaps at all are allowed.

Most noticeably, we can see that the details of the algorithm used are primarily responsible for determining the very long tracks (large zenith angles) distribution. Rejections of the multiple tracks events affect mostly those very long tracks, which are about 1% of the overall population. Then the statistics of all plots in Figure 3 is very close to the total count of 60,000 regardless.

We can now compare in detail the measured distributions obtained using the six algorithms described above, restricted to the $40° < \Theta < 70°$ region. The upper limit is based on a simple statistical argument and has practically no influence on the results of the fitting procedure. The lower limit of 40 degrees is related to the additional smoothing of the short track lengths due to the dimensions, and even the geometry, of the camera pixel matrix. An exact zenith track should give a track length of zero px, a slightly inclined track may give a track length of 1 px (or even $\sqrt{2}$), if it passes through an edge. In addition, the actual structure of the camera matrix is obviously unknown (detailed pixel shapes, spacing, etc.). All of this makes the contents of the first few bins of the path length histogram difficult to interpret and thus difficult to correct for all these effects. It should also be remembered that events in which only one pixel signal is above the threshold are rejected by the acquisition software. In the limited region, we see a near linear behavior in log × log scale, thus the power-law of the $\cos(\Theta)$ distribution. The most restrictive criteria (case Figure 3f) is determined to be the most conservative, and most suitable.

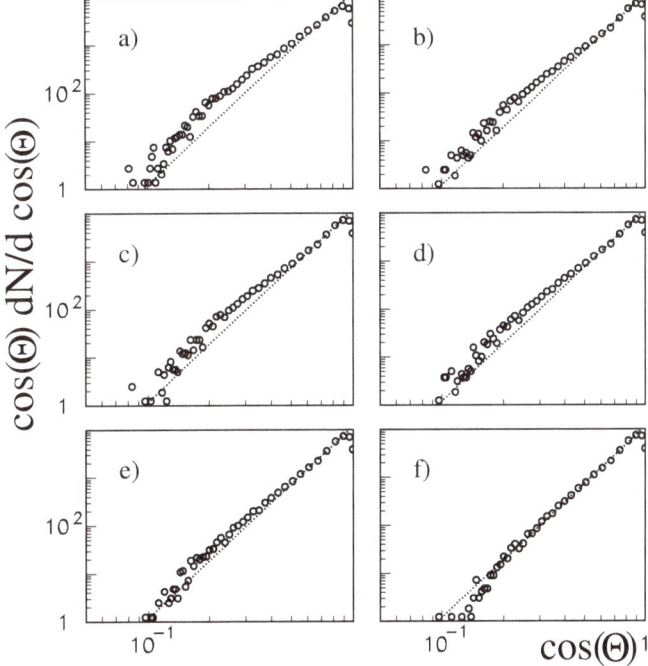

Figure 3. The observed distribution of $\cos(\Theta)$ for camera sensitive layer depth h equal to 5 px for different track length determination procedures: (**a**–**f**) (see text for detaied explanation for subfigures). The dotted line represents the expected dependency of the form $\cos^2(\Theta)$.

To be more specific, we calculate values of χ^2 comparing the measured distribution with predictions for cosmic ray muons (using $h = 5$ px). In the angular range $40° < \Theta < 70°$ there are 10 histogram bins and the respective six values are: 723, 278, 302, 237, 99, and 36. It should be remembered that these are not results of any minimization procedure, and normalizations were fixed in each case at the point of $\Theta \approx 40°$.

Henceforth, we will use the most restrictive algorithm to determine the track lengths.

3. Results

With the selected algorithm, we study the distribution of the length of the tracks recorded on smartphone photos stored in the CREDO database, in order to determine whether it is consistent with the zenith angle distribution of the muon ($\sim\cos^2(\Theta)$). The degree of the expected accordance will show the confidence of using the smartphone cameras as cosmic particle detectors.

We start detailed studies with the distribution of the track length as it is shown as the histogram in Figure 4.

Predictions assuming that we are dealing with the cosmic ray muons with the known zenith angle distribution are also shown. These distributions differ for different values of the camera photosensitive layer thickness h. The thickness of the particular camera we used in this work is not known precisely, leaving the value of h to some extent a free parameter which can be constrained by the data itself. We present here results for 3 px, 5 px, 10 px, and 15 px and the predictions of the best fitted value of h. Details of the fitting procedure are given below.

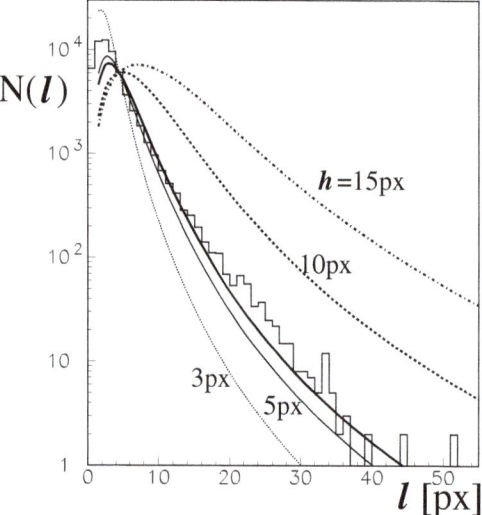

Figure 4. Measured track length distribution. Lines are shown for comparison with the predictions calculated for various values of smartphone camera matrix sensitive layer thickness (h) measured using individual pixel size (px): 3 px—dotted line, 5 px—thin continuous line, 10 px—dashed line, and 15 px—dot-dashed line. The thick solid line represents the result of our best fitted h (details in the text). On the abscissa, the value of the trace length measured, again, in the pixels size (px) is shown. All distributions are normalized at one point ($l = 5$ px).

In Figure 4, we see that the measured distribution is close to the predictions for a width (h) close to 5 px. However, it can be seen that this agreement is not perfect at both long track lengths, as well as for very short lengths. Regardless, the distribution of track length presented in Figure 4 can be converted to the zenith angle distribution (assuming we have captured cosmic ray particle traces). Zenith angle distribution is more suitable for studying cases of small angles (short trace lengths). In Figure 5, we present the comparison between the predictions obtained using different values of h.

The length of the track determined with the algorithm described above, if expressed in single pixel size (px), is, by definition, an integer number. For small values of h, which are expected in modern smartphone cameras, the measured zenith angle, given by Equation (1), would have to be discrete numbers. The resolution for small angles, at which most particles arrive at, is very limited. For larger angles, longer tracks, where the resolution of a determined angle is superior, the flux is substantially smaller, thus the simple zenith angle is not the best variable to study.

However, Figure 5 also suggests that for small angles, where most of the events occur, the matrix sensitive layer thickness is close to 5 px. While the short track discrepancy may be due to the measurement uncertainty, the discrepancy for large l values can be mitigated by selecting an appropriate value for the photosensitive layer thickness h.

The χ^2 minimization procedure was used on the $\cos(\Theta)$ distribution shown in Figure 6 where results are obtained for different assumptions about the thickness of the photosensitive layer parameter h. We have compared it with the expected power-law spectrum with a slope corresponding to the distribution of cosmic ray muons. This slope is given by the thick line in Figure 6.

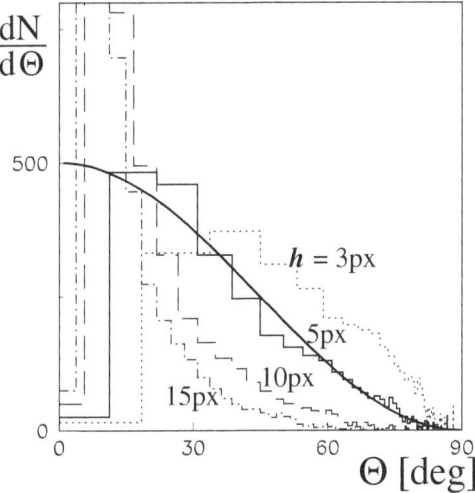

Figure 5. Zenith angle distribution obtained for various values of camera matrix thickness (h). The expected $\cos^2(\Theta)$ dependence is given by the thick line.

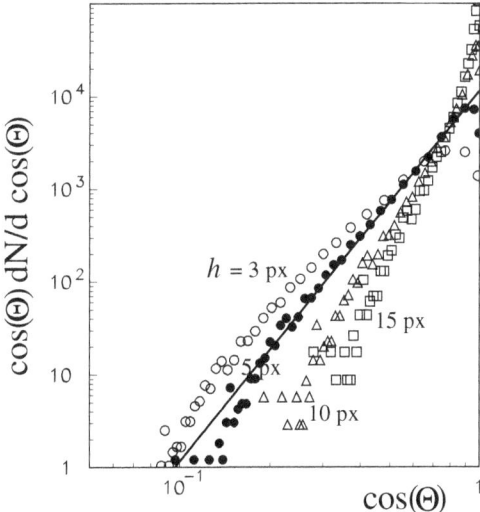

Figure 6. Observed distributions of $\cos(\Theta)$ calculated with different assumptions of the thickness of the camera matrix sensitive layer: empty circles for $h = 3$ px thickness, solid circles—5 px, triangles—10 px, and squares—15 px layer thickness. The solid line represents the expectations for the $\cos^2(\Theta)$ dependence and our best fit $h = 5.7$ px (details in the text). The right end of the abscissa corresponds to the vertical muons, while the left one to almost horizontal (89°) tracks.

The distribution of the variable $\cos(\Theta)$ illustrates very nicely the behavior of the long and very long (about horizontal) tracks. Due to the problems with short tracks we have fixed the normalization at the point corresponding to the zenith angle of ~40° and used only bins corresponding to $40° < \Theta < 70°$. The minimum value of χ^2 for 10 normal degrees of freedom was found to be 6.7 for the $h = 5.7$ px. This best solution found is shown in Figures 4 and 6 by the thick solid line.

It should be noted here that in [13,27] a similar analysis was performed and the authors concluded that the camera they used had a matrix sensitive layer thickness (termed in their work the "depletion thickness") of 29.2 ± 1.5 px, greatly exceeding our measurements. The more recent measurements of [15] have results that are compatible with a thickness of the camera matrix close to 3 px. The former analysis was based on as few as 200 registered events, the later on approximately 230 tracks. The analysis presented in this paper is based on about 60,000 tracks and hence we believe represents a significant advancement in the field.

4. Summary

We have explored the method of analyzing pictures taken by a smartphone camera with the lens obscured. In these instances, some tracks are clearly visible exceeding the dark frame camera noise. In most cases, the determination of this track length is not a significant challenge. Different methods of determining the track's main axis were tested, and lead to near indistinguishable results overall. There was some uncertainty for very long tracks corresponding to incoming particle with zenith angles greater than 80° (almost horizontal). We removed these associated ambiguities by eliminating 'long gap events', showing this ultimately to be a satisfactory solution.

The analysis of the track length distribution with the camera sensitive layer thickness considered as a 'free parameter' adjusted to the observations, agreed well with the results obtained found with a reasonable value of about 5 px (5.73 ± 0.04 px), where the unit of measure (px) is the size of the individual camera matrix pixel.

5. Conclusions

We have shown that the distribution of the zenith angle of particles responsible for the emergence of tracks in the smartphone captured images is in agreement with the expected distribution of the zenith angle of single, incoherent, cosmic ray muons. This confirms the idea that smartphones can operate in practice as 'particle pocket detectors', sensitive to charged relativistic cosmic particles and hence can be used effectively by the CREDO Project and other similar initiatives.

Author Contributions: Conceptualization, T.W.; methodology, T.W.; software, T.W.; validation, D.A.C.; formal analysis, M.K. (Michał Karbowiak); investigation, M.N.; resources, M.N. and P.H.; data curation, M.K. (Marcin Kasztelan); writing—original draft preparation, T.W.; writing—review and editing, A.R.D.; visualization, D.B.; supervision, P.H.; project administration, P.H. and D.G. All authors have read and agreed to the published version of the manuscript.

Funding: This research was partly funded by the International Visegrad grant No. 21920298.

Acknowledgments: This research has been supported in part by PLGrid Infrastructure and we warmly thank the staff at ACC Cyfronet AGH-UST for their always helpful supercomputing support. We also acknowledge the role of the users of the CREDO Detector mobile application whose data were analysed in this research.

Conflicts of Interest: The authors declare no conflict of interest.

References

1. Neronov, A.; Malyshev, D.; Semikoz, D.V. Cosmic-ray spectrum in the local Galaxy. *Astron. Astrophys.* **2017**, *606*, A22. [CrossRef]
2. Aab, A.; Abreu, P.; Aglietta, M.; Albury, J.M.; Allekotte, I.; Almela, A.; Castillo, J.A.; Alvarez-Muniz, J.; Batista, R.A.; Hahn, S.; et al. Measurement of the cosmic-ray energy spectrum above 2.5×10^{18} eV using the Pierre Auger Observatory. *Phys. Rev. D* **2020**, *102*, 062005. [CrossRef]
3. Hörandel, J.R. Models of the knee in the energy spectrum of cosmic rays. *Astropart. Phys.* **2004**, *21*, 241–265. [CrossRef]
4. Bell, A.R. Cosmic ray acceleration. *Astropart. Phys.* **2013**, *43*, 56–70. [CrossRef]
5. Zhang, Y.; Liu, S. The Origin of Cosmic Rays from Supernova Remnants. *Chin. Astron. Astrophys.* **2020**, *44*, 1–31.

6. Batista, R.A.; Biteau, J.; Bustamante, M.; Dolag, K; Engel, R.; Fang, K; Kampert, K.-H., Kostunin, D.; Mostafa, M; Murase, K.; et al. Open Questions in Cosmic-Ray Research at Ultrahigh Energies. *Front. Astron. Space Sci.* **2019**, *6*, 23. [CrossRef]
7. Góra, D.; Almeida Cheminant, C.; Alvarez-Castillo, D.; Bratek, Ł; Dhital, N.; Duffy, A.R.; Homola, P.; Jagoda, P.; Jałocha, J.; Kasztelan, M.; et al. Cosmic-Ray Extremely Distributed Observatory: Status and Perspectives. *Universe* **2018**, *4*, 2218.
8. Homola, P.; Bhatta, G.; Bratek, Ł.; Bretz, T.; Almeida Cheminant, K.; Alvarez-Castillo, D.; Dhital, N.; Devine, J.; Góra, D.; Jagoda, P.; et al. Search for Extensive Photon Cascades with the Cosmic-Ray Extremely Distributed Observatory. *CERN Proc.* **2018**, *1*, 289.
9. Woźniak K.W.; Almeida-Cheminant, K.; Bratek, Ł.; Alvarez Castillo, D.E.; Dhital, N.; Duffy, A.R.; Góra, D.; Hnatyk, B.; Homola, P.; Jagoda, P.; et al. Detection of Cosmic-Ray Ensembles with CREDO. *EPJ Web Conf.* **2019**, *208*, 15006.
10. Bibrzycki, Ł.; Burakowski, D.; Homola, P.; Piekarczyk, M.; Niedźwiecki, M.; Rzecki, K.; Stuglik, S.; Tursunov, A.; Hnatyk, B.; Castillo; et al. Towards A Global Cosmic Ray Sensor Network: CREDO Detector as the First Open-Source Mobile Application Enabling Detection of Penetrating Radiation. *Symmetry* **2020**, *12*, 1802. [CrossRef]
11. Whiteson, D.; Mulhearn, M.; Shimmin, C.; Cranmer, K.; Brodie, K.; Burns, D. Searching for ultra-high energy cosmic rays with smartphones. *Astropart. Phys.* **2016**, *79*, 1. [CrossRef]
12. Meehan, M.; Bravo, S.; Campos, F.; Ruggles, T.; Schneider, C.; Vandenbroucke, J.; Winter, M. The particle detector in your pocket: The Distributed Electronic Cosmic-ray Observatory. In Proceedings of the ICRC 2017, Busan, Korea, 10–20 July 2017; p. 375.
13. Vandenbroucke, J.; Bravo, S.; Karn, P.; Meehan, M.; Plewa, M.; Ruggles, T.; Schultz, D.; Peacock, J.; Simons, A.L. Detecting particles with cell phones: the Distributed Electronic Cosmic-ray Observatory. In Proceedings of the ICRC 2015, Hague, The Netherlands, 30 July–6 August 2015; p. 691.
14. Borisyak, M.; Usvyatsov. M.; Mulhearn, M.; Shimmin, C.; Ustyuzhanin, A. Muon Trigger for Mobile Phones. *J. Phys. Conf. Ser.* **2017**, *898*, 032048. [CrossRef]
15. Davoudifar, P.; Bagheri, Z. Determination of local muon flux using astronomical Charge Coupled Device., *J. Phys. Nucl. Part. Phys.* **2020**, *47*, 035204. [CrossRef]
16. Groom, D. Cosmic rays and other nonsense in astronomical CCD imagers. *Exp. Astron.* **2002**, *14*, 45. [CrossRef]
17. Unger, M.; Farrar, G. (In)Feasability of Studying Ultra-High-Energy Cosmic Rays with Smartphones. *arXiv* 2015, arXiv:1505.04777.
18. Antoni, T.; Apel, W.D; Badea, F.; Bekk, K.; Bercuci, A.; Blumer, H.; Bozdog, H.; Brancus, I.M.; Buttner, C; Chilingarian, A.; et al. The cosmic-ray experiment KASCADE. *Nucl. Instrum. Methods Phys. Res. A* **2003**, *513*, 490. [CrossRef]
19. Ogio, S.; Kakimoto, F.; Kurashina, Y.; Burgoa, O.; Harada, D.; Tokuno, H.; Yoshii, H.; Morizawa, A.; Gotoh, E.; Nakatani, H.; et al. The Energy Spectrum and the Chemical Composition of Primary Cosmic Rays with Energies from 10^{14} to 10^{16} eV. *Astrophys. J.* **2004**, *612*, 268. [CrossRef]
20. Gupta, S.K.; Aikawa, Y.; Gopalakrishnan, N.V.; Hayashi, Y.; Ikeda, N.; Ito, N.; Jain, A.; John, A.V.; Karthikeyan, S.; Kawakami S.; et al. GRAPES-3—A high-density air shower array for studies on the structure in the cosmic-ray energy spectrum near the knee. *Nucl. Instrum. Methods Phys. Res. A* **2005**, *540*, 311. [CrossRef]
21. Wang, D.-R.; Wang, X.-P.; Li, C.-Z.; Chen, Y.-B.; Lin, J.-F.; Chang, C.-C.; Zhao, J.-W.; Jiang, Y.-Y.; Xu, Y.-L.; Tang, S.-Q.; et al.Tthe measurement of the cosmic ray muon zenith angle distribution with on-line microcomputer. *Chin. Phys. C* **1983**, *7*, 135.
22. Franke, R.; Holler, M.; Kaminsky, B.; Karg, T.; Prokoph, H.; Schönwald, A.; Schwerdt, C; Stößl, A.; Walter, M. CosMO—A Cosmic Muon Observer Experiment for Students. In Proceedings of the ICRC2013, Rio de Janeiro, Brazil, 2–9 July 2013; p. 1084.
23. Hutten, M.; Karg, T.; Schwerdt, C.; Steppa, C.; Walter, M. The International Cosmic Day—An Outreach Event for Astroparticle Physics. In Proceedings of the ICRC 2017, Busan, Korea, 10–20 July 2017; p. 406.
24. Singh, P.; Hedgel, H. Special relativity in the school laboratory: a simple apparatus for cosmic-ray muon detection. *Phys. Educ.* **2015**, *50*, 317. [CrossRef]
25. Greisen, K. Intensity of Cosmic Rays at Low Altitude and the Origin of the Soft Component. *Phys. Rev.* **1943**, *63*, 323. [CrossRef]
26. Rossi, B. Interpretation of Cosmic-Ray Phenomena. *Rev. Mod. Phys.* **1948**, *20*, 537. [CrossRef]
27. Vandenbroucke, J.; BenZvi, S.; Bravo, S.; Jensen, K.; Karn, P.; Meehan, M.; Peacock, J.; Plewa, M.; Ruggles, T.; Santander, M. et al. Measurement of cosmic-ray muons with the Distributed Electronic Cosmic-ray Observatory, a network of smartphones. *J. Instrum.* **2016**, *11*, P04019. [CrossRef]

Article

Storm-Time Features of the Ionospheric ELF/VLF Waves and Energetic Electron Fluxes Revealed by the China Seismo-Electromagnetic Satellite

Zeren Zhima [1,*], Yunpeng Hu [2], Xuhui Shen [1], Wei Chu [1], Mirko Piersanti [3], Alexandra Parmentier [4], Zhenxia Zhang [1], Qiao Wang [1], Jianping Huang [1], Shufan Zhao [1], Yanyan Yang [1], Dehe Yang [1], Xiaoying Sun [1], Qiao Tan [1], Na Zhou [1] and Feng Guo [1]

[1] National Institute of Natural Hazards, MEMC, Beijing 100085, China; xuhuishen@ninhm.ac.cn (X.S.); weichu@ninhm.ac.cn (W.C.); zhenxiazhang@ninhm.ac.cn (Z.Z.); qiaowang@ninhm.ac.cn (Q.W.); jianpinghuang@ninhm.ac.cn (J.H.); shufanzhao@ninhm.ac.cn (S.Z.); yanyanyang@ninhm.ac.cn (Y.Y.); deheyang@ninhm.ac.cn (D.Y.); xiaoyingsun@ninhm.ac.cn (X.S.); qiaotan@ninhm.ac.cn (Q.T.); nazhou@ninhm.ac.cn (N.Z.); fengguo@ninhm.ac.cn (F.G.)
[2] School of Space and Environment, Beihang University, Beijing 100191, China; Huyunpeng15@163.com
[3] INAF-Istituto di Astrofisica e Planetologia Spaziali, 00133 Rome, Italy; mirko.piersanti@roma2.infn.it
[4] National Institute of Nuclear Physics, Division of Rome "Tor Vergata", 00186 Rome, Italy; parmentier@roma2.infn.it
* Correspondence: zerenzhima@ninhm.ac.cn

Abstract: This study reports the temporal and spatial distributions of the extremely/very low frequency (ELF/VLF) wave activities and the energetic electron fluxes in the ionosphere during an intense storm (geomagnetic activity index Dst of approximately −174 nT) that occurred on 26 August 2018, based on the observations by a set of detectors onboard the China Seismo-Electromagnetic Satellite (CSES). A good correlation of the ionospheric ELF/VLF wave activities with energetic electron precipitations during the various storm evolution phases was revealed. The strongest ELF/VLF emissions at a broad frequency band extending up to 20 kHz occurred from the near-end main phase to the early recovery phase of the storm, while the wave activities mainly appeared at the frequency range below 6 kHz during other phases. Variations in the precipitating fluxes were also spotted in correspondence with changing geomagnetic activity, with the max values primarily appearing outside of the plasmapause during active conditions. The energetic electrons at energies below 1.5 MeV got strong enhancements during the whole storm time on both the day and night side. Examinations of the half-orbit data showed that under the quiet condition, the CSES was able to depict the outer/inner radiation belt as well as the slot region well, whereas under disturbed conditions, such regions became less sharply defined. The regions poleward from geomagnetic latitudes over 50° were found to host the most robust electron precipitation regardless of the quiet or active conditions, and in the equatorward regions below 30°, flux enhancements were mainly observed during storm time and only occasionally in quiet time. The nightside ionosphere also showed remarkable temporal variability along with the storm evolution process but with relatively weaker wave activities and similar level of fluxes enhancement compared to the ones in the dayside ionosphere. The ELF/VLF whistler-mode waves recorded by the CSES mainly included structure-less VLF waves, structured VLF quasi-periodic emissions, and structure-less ELF hiss waves. A wave vector analysis showed that during storm time, these ELF/VLF whistler-mode waves obliquely propagated, mostly likely from the radiation belt toward the Earth direction. We suggest that energetic electrons in the high latitude ionosphere are most likely transported from the outer radiation belt as a consequence of their interactions with ELF/VLF waves.

Keywords: ionosphere; ELF/VLF waves; energetic electron precipitations; storm-time feature; CSES

Citation: Zhima, Z.; Hu, Y.; Shen, X.; Chu, W.; Piersanti, M.; Parmentier, A.; Zhang, Z.; Wang, Q.; Huang, J.; Zhao, S.; et al. Storm-Time Features of the Ionospheric ELF/VLF Waves and Energetic Electron Fluxes Revealed by the China Seismo-Electromagnetic Satellite. *Appl. Sci.* **2021**, *11*, 2617. https://doi.org/10.3390/app11062617

Academic Editor: Hyung-Sup Jung

Received: 12 February 2021
Accepted: 8 March 2021
Published: 15 March 2021

Publisher's Note: MDPI stays neutral with regard to jurisdictional claims in published maps and institutional affiliations.

Copyright: © 2021 by the authors. Licensee MDPI, Basel, Switzerland. This article is an open access article distributed under the terms and conditions of the Creative Commons Attribution (CC BY) license (https://creativecommons.org/licenses/by/4.0/).

1. Introduction

Within the solar-terrestrial system, the ionospheric layers can mirror both Earthward and Sunward disturbances at various scales. From the Earth direction, namely from the lithosphere or atmosphere, the disturbances include the relatively weak but detectable seismic precursor anomalies [1], the strong radio waves emitted by powerful ground-based very low frequency (VLF) transmitters [2], the harmonic radiation from the electric power lines [3], and the strong whistler-mode waves induced by lightning/thunderstorms in the atmosphere. From the Sun direction, the disturbances coming from above the ionosphere are mainly related to the Sun, which releases coronal mass ejections (CMEs) and solar flares, as well as the resultant geomagnetic storm/substorm activity. In particular, there is a magnetic field-line mapping of the magnetosphere onto the high-latitude ionosphere, along which field-line currents, possibly associated with magnetospheric polarization fronts, enter the post-midnight ionosphere and leave it from the pre-midnight side, thus feeding the strong aurora electrojet [4,5]. This makes the ionosphere a highly dynamic system where a variety of intense electromagnetic emissions and energetic particle precipitations are detected due to multiple layer interactions with the lithosphere, atmosphere, and magnetosphere, as well as solar wind [4–8].

The electromagnetic emissions excited in the extremely/very low frequency (ELF/VLF) range [9–13] are the most direct outcomes of this dynamic system. The most commonly observed typical electromagnetic waves at ELF/VLF bands are the whistler-mode waves [9,14–18], which include the whistler-mode VLF chorus [12,16,19], quasi-periodic waves [13,15,20], ELF hiss [21–23], and strong whistlers induced by lightning, VLF radio wave transmitters, or other sources [22,24]. These typical whistler-mode waves play significant roles either in the acceleration or loss process of relativistic electrons in the Earth's outer radiation belt [14,17,25]. The VLF chorus mainly appears at frequency from 0.1 to 0.8 f_{ce} (the equatorial electron cyclotron frequency) and can efficiently accelerate relativistic electrons (E > MeV) [14]. Zhima et al. [10] firstly found the evidence of whistler-mode chorus penetrating the plasmapause and entering into the low altitude ionosphere, and Zhang et al. [26] firstly reported that chorus waves in the magnetosphere accelerated energetic electrons (1–3 MeV) in the ionosphere from the China Seismo-Electromagnetic Satellite (CSES) observations. ELF hiss waves are structure-less and incoherent electromagnetic waves that preferentially appear at a broad frequency range from several hundred Hz to 3 kHz, playing vital roles in either the loss of energetic electrons or the formation of radiation belt slot region [22,27,28]. Another important ELF/VLF whistler-mode wave often appearing in the ionosphere is the quasi-periodic (QP) wave at frequencies from several hundred Hz to ~4 kHz with varying periodic modulations of wave intensity over time scales from several seconds to a few minutes; they are often observed by low Earth orbit (LEO) satellites and ground stations [13,15]. Zhima et al. [20] reported the well-pronounced rising-tone structure QP waves and simultaneous energetic electron precipitations (from ~400 keV to 1 MeV) in the high-latitude ionosphere firstly based on the CSES's observations.

Because they are simultaneously observed with the strong ELF/VLF wave activity, changes in the fluxes of energetic particle populations are usually checked for. In the context of radiation-belt dynamics, it is widely accepted that ELF/VLF whistler-mode waves represent the main candidate for the acceleration/loss of particles during storm time [17,29], so they have been given increasing attention due to their impact on the dynamics of near-Earth space. Benck et al. [18] revealed the global distribution of ionospheric ELF/VLF wave intensities and energetic electron precipitations during an intense magnetic storm by making use of DEMETER (Detection of Electromagnetic Emissions Transmitted from Earthquake Regions) observations (altitude: ~710 km in 2004), and a good correlation between waves and particle precipitations at different stages of storm evolution was revealed. Still relying on DEMETER observations, Zhima et al. [9] statistically analyzed the temporal–spatial variations of the ionospheric ELF/VLF waves during all the intense CME-driven storms that occurred from 2005 to 2009 (altitude: ~660 km), specifically showing how the ELF/VLF waves in different frequency ranges were excited across the L shell space during the evolution of this type of storms.

Since the successful operation of the French DEMETER satellite (2005–2010), which was mainly dedicated to the investigation of ionospheric disturbances possibly associated with the strong earthquakes, volcanoes, or anthropogenic activities from the lithosphere [30], a growing number of studies [8,11,31,32] obtained from DEMETER observations have been suggesting that satellites probing electromagnetism in LEO space can be regarded as promising tools to monitor natural hazards [33].

However, since DEMETER was decommissioned in December 2010, there lacks similar type of electromagnetic satellite in the ionosphere to provide first-hand observations for studying natural disasters until the recent successful launch of the CSES on 2 February 2018 at an altitude of 507 km in the ionosphere [33,34]. The CSES is part of China's Zhangheng mission, which is aimed to launch both electromagnetic and gravity satellites in near-Earth orbit within the next few decades. The Zhangheng mission is named after the ancient scientist Zhangheng who invented the world's first seismoscope in the second century CE. The Zhangheng-01 (ZH-1) mission is dedicated to the electromagnetic satellites with three planned consecutive launches—it is also known as the CSES; Zhangheng-02, in the preparation phase at present, is dedicated to gravity satellites.

The first probe of the CSES mission was successfully launched into a sun-synchronous circular orbit. The second probe will be launched into the same orbit before December 2022, but it will operate at the opposite side of the Earth, which means that two probes will be operating on both the day and night sides of the ionosphere by 2023 (the first probe was designed to span a five-year lifetime but is expected to operate longer).

Since its successful launch in 2018, the CSES has been providing valuable measurements about the electromagnetic field across a broad frequency range from ultra-low frequency (ULF) to very low frequency (VLF), and even to the high frequency (HF) of the electric field, as well as fluxes of energetic electrons at energies from 0.1 to 50 MeV. In view of the importance of electromagnetic wave activities and energetic electron populations for natural hazard science, the purpose of this paper was to investigate the features of waves and particles in LEO space by using CSES observations. Based on previous DEMETER studies [9,18] (at altitudes from ~660 to 710 km), this work presents CSES data to describe the temporal and spatial distribution of the ionospheric ELF/VLF waves and energetic particle fluxes during the G3-class storm (Dst minimum of -174 nT) that occurred on 26 August 2018. A brief introduction to the CSES and associated payloads is provided in Section 2, while the distributions of ionospheric ELF/VLF waves and energetic electron fluxes are presented in Section 3. Section 4 offers a discussion concerning experimental observations, and Section 5 briefly summarizes the main results.

2. Satellite and Data

The main scientific objective of the CSES is to monitor ionospheric perturbations associated with natural hazards (mainly with strong earthquakes) in the quest for possible earthquake forecasting. To serve the need of emergence response to disastrous earthquakes, such as the 2008 Mw 7.9 Wenchuan earthquake, the CSES was designed to provide real-time data over China's territory via a direct downlink to the ground segment.

The CSES completes 15.2 orbits around Earth per day, with an orbital period of ~94.6 min and a five-day recursive period over the same geographic area with the ascending/descending node local time of 02 a.m./02 p.m., respectively. Its payloads allow for the measurement of the electromagnetic field, energetic particles, and in-situ and profile ionosphere parameters. The diverse physical parameters obtained by the CSES can support the comprehensive research of geo-/space physics and radio science.

The payloads involved in this study are briefly introduced here. The background geomagnetic field is measured by a high precision magnetometer (HPM) [35], which includes two fluxgate sensors [36] and a coupled dark-state magnetometer [37], providing both vector and scalar values of the total geomagnetic field in a frequency range from DC to 15 Hz and a sampling rate of 1 Hz for the scalar geomagnetic field detection. The variant magnetic field from 10 Hz to 20 kHz is measured by a tri-axis search-coil magnetometer (SCM) mounted on the end part of a 4.5-m-long boom to avoid the artificial electromagnetic interferences induced by the satellite platform itself [38,39]. The SCM provides data in the three frequencies bands of ULF (10 to 200 Hz), ELF (from 200 Hz to 2.2 kHz), and VLF (1.8 to 20 kHz) with sampling rates of 1024 Hz, 10.24 kHz, and 51.2 kHz, respectively. The electric field is detected by an electric field detector (EFD), which consists of four spherical sensors mounted at the near-end part of four booms (4.5 m long) [40], measuring the electric field in four frequency channels: ULF (DC to16 Hz), ELF (from 6 Hz to 2.2 kHz), VLF (1.8 to 20 kHz), and HF (from 18 kHz to 3.5 MHz), with sampling rates of 128 Hz, 5 kHz, 50 kHz, and 10 MHz, respectively. The energetic particles are recorded by the high energetic particle package (HEPP), with three sub-detectors (HEPP-H, HEPP-L, and HEPP-X) providing the energy spectrum and pitch-angle distribution of charged particles (protons: from 2 MeV to 200 MeV; electrons: from 100 keV to 50 MeV) and soft X-ray emission from solar events [41,42]. In this study, HEPP-L was selected to measure electron fluxes at energies from 0.1 to 3 MeV and corresponding pitch-angle distributions. HEPP-L contains nine silicon-slice units and one anti-coincidence detector with an angular resolution of 5° and a maximum field-view of 100° × 30° [42].

3. Observations

Figure 1 shows major parameters describing the solar wind conditions and the geomagnetic activity indices Dst and Kp from 20 August to 4 September 2018. It can be seen that the north–south component of the interplanetary magnetic field (IMF Bz) in the GSM (geocentric solar magnetospheric) coordinate system suffered a slight northern enhancement of up to ~9 nT at ~14:00 UT on 25 August 2018 before suddenly turning southward with a steep drop to the minimum value of −16.8 nT at 05:00 UT on 26 August, followed by a very fluctuating recovery back to an average value of ~0 nT. This prolonged southward alignment of the IMF Bz triggered a dayside magnetic field line reconnection at the magnetopause, directly inducing a strong geomagnetic storm during which the geomagnetic equatorial Dst index dropped to the minimum value of −174 nT at 06:00 to 07:00 UT on 26 August, followed by a long-term recovery phase lasting until 3 September 2018. The source for this intense geomagnetic storm was identified as the interplanetary coronal mass ejection (ICME) from the Sun that occurred downstream of a filament eruption on 20 August 2018 [43]. In particular, from the point of view of solar wind velocity, this ICME event was pretty interesting, since a fast speed of more than 600 km/s, along with the subsequent consistent slowdown to less than 400 km/s and slight recovery, was reached prior to the onset of the geomagnetic storm before again giving rise to a fast speed (up to ~500 km/s) in the early recovery phase and a new marked decrease in the late part of the recovery. The solar wind dynamic pressure, which was a direct indicator of solar wind and ion and electron density, intensely increased during the main and early recovery phases, reached the peak at the early recovery phase, and remained fluctuating around 2 nPa for the most of other times.

The effects of this geomagnetic storm on Earth's ionosphere were investigated by Younas et al. [44] based on multi-parameter measurements, including total electron content (TEC), geomagnetic field intensity, and O/N2 ratio data, and theit results showed that a positive ionosphere storm (increase of TEC) in the southern hemisphere and a negative storm in the northern one were formed. Yang et al. [5] also reported simultaneous responses from multi-type CSES payloads to this event, thus leading to the assessment of the good performance of the CSES in reacting to the different phases of the storm. Strong ELF/VLF emissions were especially observed, with the simultaneous enhancement of energetic electron fluxes in the South Atlantic Anomaly (SAA), outer/inner radiation belt, and slot region [5].

In the following section, we mainly focus on the temporal and spatial distributions of the storm-time ELF/VLF waves across a broad frequency band from 200 Hz to 20 kHz, as well as the energetic electron fluxes at energies 0.1 MeV < E < 3.0 MeV by using data from the SCM, EFD, and HEPP-L onboard the CSES.

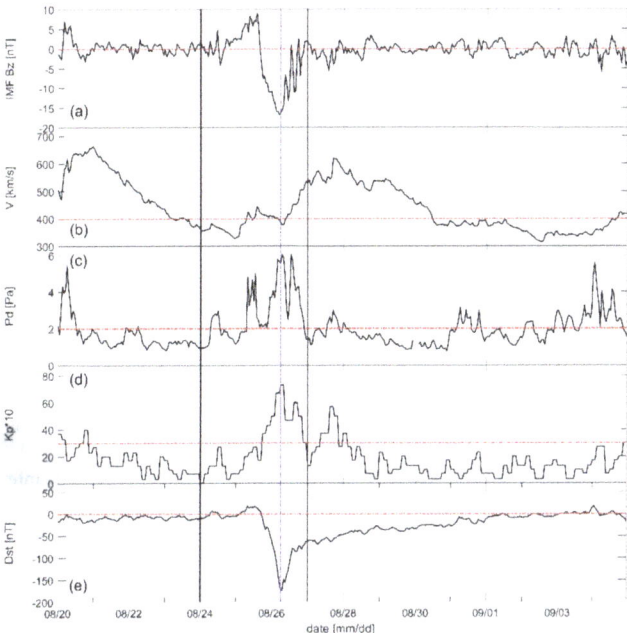

Figure 1. The variation of solar wind conditions and geomagnetic Dst and Kp indices from 20 August to 4 September 2018. From top to bottom: (**a**) the z component of the interplanetary magnetic field in the geocentric solar magnetospheric (GSM) coordinate system (interplanetary magnetic field (IMF) Bz), (**b**) solar wind velocity, (**c**) solar dynamic pressure, (**d**) geomagnetic activity Kp index, and (**e**) Dst index.

We adopted the superposed epoch analysis method by defining the reference time t_0 as the time of Dst minimum (from 06:00 UT on 26 August 2018), as we did in previous work [6]. We mainly computed the average power spectral density (PSD) values of the electric and magnetic field at a frequency range of 1 kHz < f < 2 kHz, as well as the average energetic electron fluxes data at an energy level of 1 MeV < E < 3 MeV for each orbit. Then, these data were binned as a function of L shell in steps of 0.2 L, with a time interval of 1 h in the epoch time period from $t_0 - 30$ h to $t_0 + 90$ h (namely, from 00:00 UT 25 August to 00:00 UT 30 August).

The results are presented in Figure 2 (dayside) and Figure 3 (nightside). As seen in Figure 2c,d, before the storm initial phase that began at epoch time $t_0 - 16$ h (14:00 UT on 25 August 2018) and indicated by a slight increase of Dst index, the ionosphere remained in a relatively quiet electromagnetic environment (denoted by the blue areas in Figure 2c,d from epoch time $t = -30$ h to -16 h). Meanwhile, along with the consistent southern reversal of IMF Bz and the corresponding decrease of the Dst index, the wave activities started to grow up (~$t = -9$ h) and reached ELF/VLF maximum emissions around the end of the main phase and the beginning of the recovery phase ($t = -2$ to 20 h). We can see a close correlation between the ELF/VLF wave activities, the IMF Bz, and the solar wind (velocity and pressure); when IMF Bz and solar wind were stable at times when IMF Bz was close to 0 nT, the wave activities were weak (e.g., 20 h < t < 28 h and 60 h < t < 72 h), while when the solar wind conditions were strongly fluctuating, strong ELF/VLF emissions were observed (e.g., 2 h < t < 20 h).

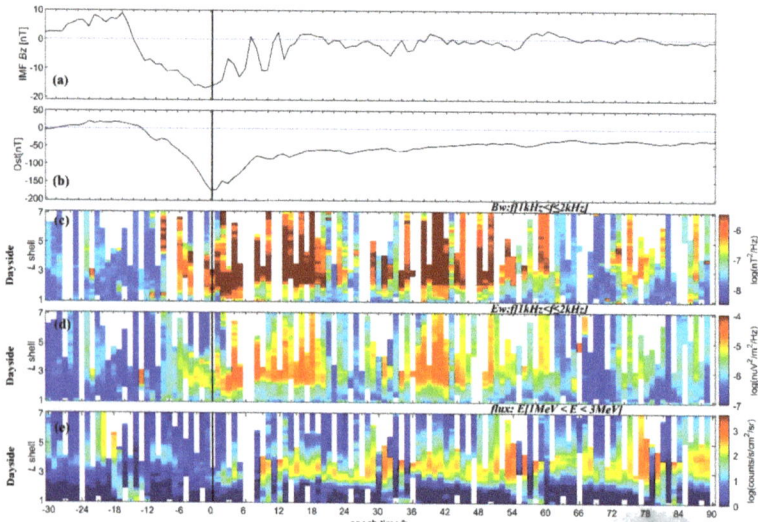

Figure 2. The variation of ionospheric extremely/very low frequency (ELF/VLF) wave intensity and energetic electron fluxes during the geomagnetic storm that occurred from 25 to 29 August 2018. (**a**) IMF Bz in the GSM coordinates, (**b**) Dst index, (**c,d**) the power spectral density (PSD) values of the magnetic and the electric field at frequency range from 1 to 2 kHz, and (**e**) the energetic electron fluxes at energy level from 1 to 3 MeV. Data were integrated by the superposed epoch method [9], and the vertical solid line denotes the reference time t_0 that is the time of Dst minimum on 06:00 to 07:00 UT on 26 August 2018. The y axis in (**c–e**) indicates the L shell from 1 to 7. The white blank regions denote data gaps.

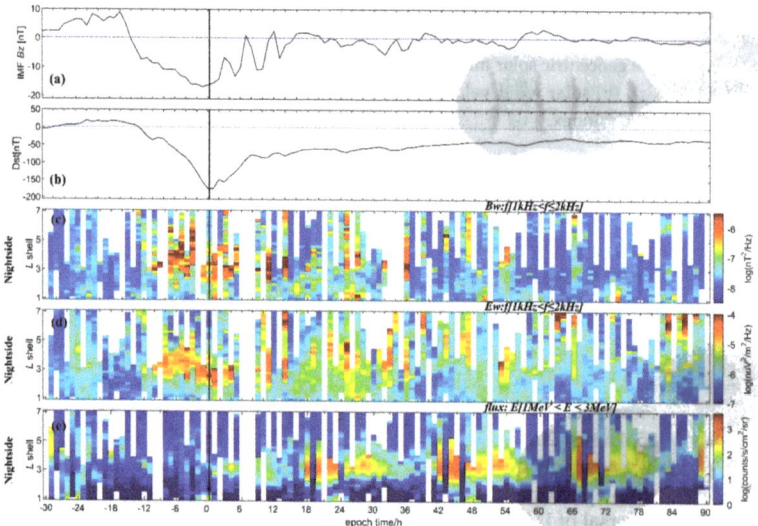

Figure 3. Same as Figure 2 but for the nightside ionosphere.

Figure 2e shows that the energetic electron populations at 1 MeV < E < 3 MeV were also smaller before the initial and early main phases but gradually increased to a higher level during the near-end main phase and remained to be enhanced during the whole recovery phase. The energetic electron populations were stronger at L shell between ~2 and 3, where it was expected to find the inner radiation belt [26]. Figure 2 shows that both waves and fluxes clearly reflected the remarkable temporal variability of the boundary of the inner radiation belt (i.e., filling and fluctuations of the magnetosphere slot region).

Figure 3 shows the results for the nightside ionosphere, where the ELF/VLF wave activities and electron fluxes also showed a close correlation of the different storm phases, though with a relatively weaker intensity than the dayside. It is interesting that in the nightside ionosphere, the waves at 1 kHz < f < 2 kHz were predominantly enhanced during the main phase in contrast to the ones at dayside; this feature is worth further exploration after more storm cases are accumulated.

To further depict the temporal and spatial distribution of the wave activities of all the ELF/VLF frequency bands, we divided the electromagnetic field measurements into seven frequency bands: 0.2–1 kHz, 1–3 kHz, 3–6 kHz, 6–9 kHz, 9–12 kHz, 12–15 kHz, and 15–20 kHz. By using the same method from Figure 2, the average PSD values of each band from each orbit were computed and binned as a function of L shell values in a step of 0.2 L, with a time interval of 1 h in the epoch time period from $t_0 - 30$ h to $t_0 + 90$ h. Similar for the energetic electrons, the fluxes from the six energy bands were integrated: 0.1–0.5 MeV, 0.5–1 MeV, 1.0–1.5 MeV, 1.5–2.0 MeV, 2–2.5 MeV, and 2.5–3 MeV.

Figures 4 and 5 show the results of ELF/VLF waves in the magnetic field from the day and night side ionospheres, respectively. In Figures 4 and 5, the black vertical dashed lines denote the reference epoch time t_0. It can be seen that the arrival of an ICME at the magnetopause location directly led to strong ELF/VLF wave activities in the ionosphere. Specifically, before the ICME hit the magnetopause ($t < t_0 - 16$ h), the ELF/VLF wave activities at all frequency bands were relatively quiet. At the early stage of main phase (from $t_0 - 16$ h to ~$t_0 - 4$ h), the ELF/VLF waves were mainly excited at frequencies below 6 kHz. However, from the near-end main phase to the early recovery phase (from $t_0 - 6$ h to ~$t_0 + 6$ h), the wave activities significantly grew up in the entire frequency range from 0.2 to 20 kHz. During the long recovery phase, the wave enhancements mainly occurred below 6 kHz. These results from the CSES were consistent with our previous statistical analysis for CME-driven storms based on DEMETER observations [9].

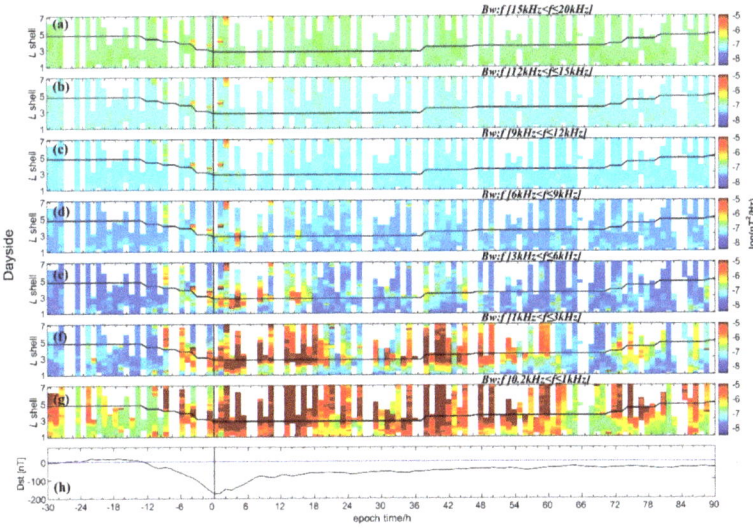

Figure 4. The temporal and spatial evolution of the ionospheric ELF/VLF wave intensities at frequency range from 0.2 to 20 kHz at the magnetic field in the dayside ionosphere, revealed by the search-coil magnetometer (SCM) onboard the China Seismo-Electromagnetic Satellite (CSES) during the geomagnetic storm on 26 August 2018. From top to bottom: the PSD values at different frequency bands of (**a**) 15–20 kHz, (**b**) 12–15 kHz, (**c**) 9–12 kHz, (**d**) 6–9 kHz, (**e**) 3–6 kHz, (**f**) 1–3 kHz, and (**g**) 0.2–1 kHz. The overlapping black curves denote the location of plasmapause estimated by the (**h**) Dst index. The vertical dashed line denotes the time of the Dst minimum.

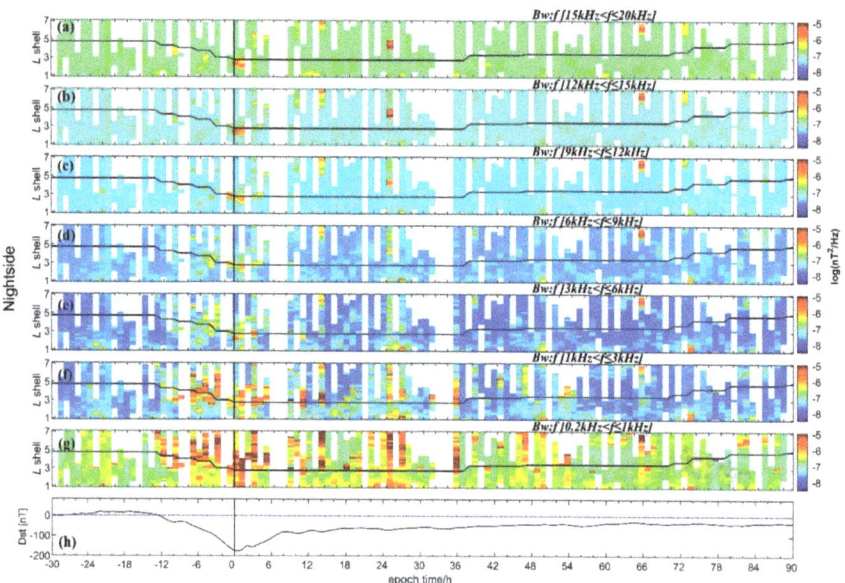

Figure 5. Same as Figure 4 but for the ELF/VLF wave intensities of the magnetic field in the nightside ionosphere.

Figure 5 shows that the nightside wave activities were weaker than those on the dayside. They also predominantly appeared at frequency below 3 kHz, and the phenomenon of the whole frequency band enhancement only appeared during the hours when the Dst reached its minimum values. The results of the electric field are not presented here, partially because they showed very similar features to those seen in the magnetic field and partially because there were high noises over the equatorial area that originated from instruments or satellite platforms (see the strong horizontal enhancement denoted by black rectangles in Figures 8 and 9 in Discussion Section).

We also estimated the location of the plasmapause by using the empirical linear model of O'Brien and Moldwin [45], which was based on the magnetic indices Kp and Dst values. According to that model, the plasmapause (denoted by the black curves superposed on the plots in Figures 4 and 5) was significantly compressed during the storm time from L shell values of ~5 to ~3. According to the estimated plasmapause location, it can be said that the strong wave emissions at the frequencies below 6 kHz mainly occurred at a broad L shell extension (covering the L shell from ~1 to 7), both inside and outside the plasmapause; meanwhile, waves at frequencies higher than 9 kHz mainly appeared outside the plasmapause (L shell ~3 to 7), which means that waves in this portion of the spectrum mainly appeared in the radiation belt (more likely in the outer radiation belt).

Figures 6 and 7 show the temporal and spatial evolution of energetic electron fluxes at energies from 0.1 to 3 MeV in the day and night side ionospheres, respectively. During the main phase of the storm, the electron flux at $E < 1$ MeV intensely increased, but the ones at higher energy level $E > 1.5$ MeV decreased (see $t = -12$ h to 0 h). Subsequently, during the recovery phase of the storm, when the solar wind parameters and the Dst index were returning back to their nominal level, the fluxes at $E > 1.5$ MeV got enhanced to about 2 orders higher than those at the pre-storm values. Features in electron flux variations during storm evolution were consistent with earlier DEMETER observations in storm time [18].

Clearly, the precipitating flux increased with the geomagnetic activity level, with the primary maximization out of the plasmapause under active conditions. Specifically, sub-MeV (below $E > 1.5$) electrons showed strong enhancements across the whole storm time on both the day and night sides, as well as both inside and outside the plasmapause; however, for $E > 1.5$ MeV, the fluxes were mainly enhanced over the plasmapause.

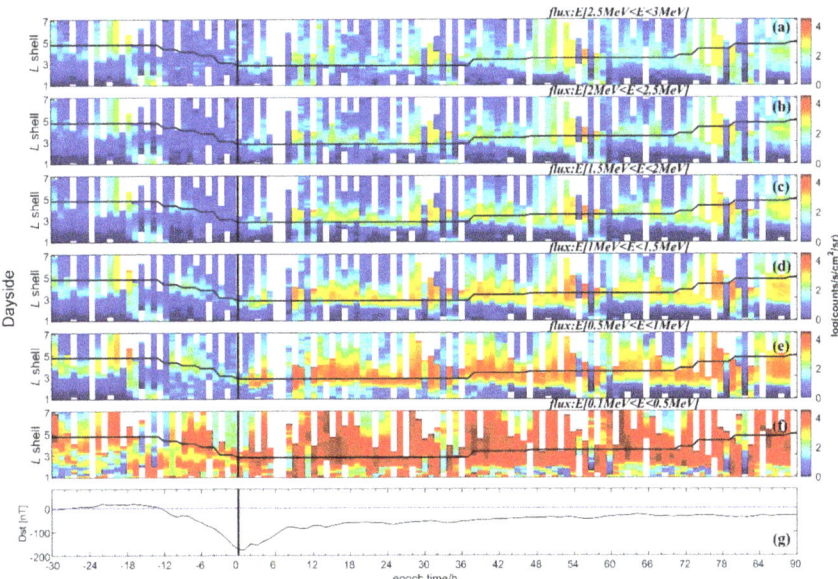

Figure 6. The dayside temporal and spatial evolution of energetic electron fluxes at energies from 0.1 to 3 MeV detected by high energetic particle package (HEPP)-L on board the CSES during the storm that occurred on 26 August 2018. From top to bottom: (**a**) 2.5–3 MeV, (**b**) 2–2.5 MeV, (**c**) 1.5–2.0 MeV, (**d**) 1.0–1.5 MeV, (**e**) 0.5–1 MeV, and (**f**) 0.1–0.5 MeV; (**g**) the Dst index. The overlapping black curves denote the location of the plasmapause estimated from Dst values. The vertical dashed line marks the time of the Dst minimum.

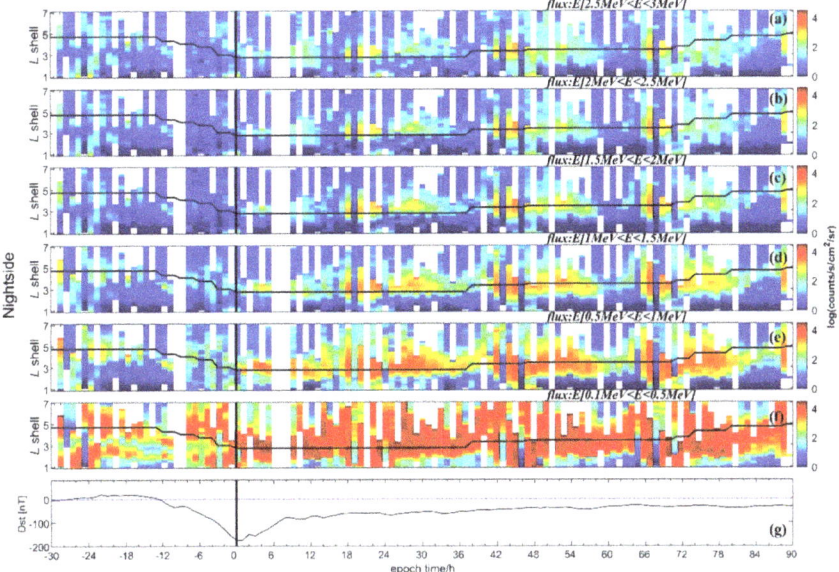

Figure 7. Same as Figure 6 but for the nightside ionosphere.

4. Discussion

We present two half-orbit observations from the CSES as examples to discuss the exact electromagnetic environment in the ionosphere under geomagnetic quiet (Figure 8) and disturbed conditions (Figure 9).

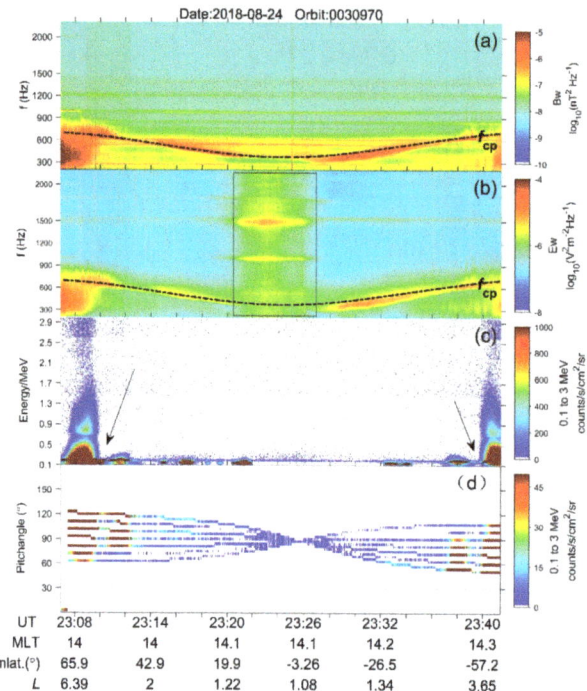

Figure 8. The ELF/VLF wave activities and energetic electron fluxes on the dayside ionosphere during magnetic quiet time recorded by No. 0030970 on 24 August 2018. From top to bottom: (**a**,**b**) power spectral density values of the magnetic and electric fields, (**c**) the energetic fluxes at energy band from 0.1 to 3.0 MeV, and (**d**) the energetic electron fluxes distributions along the pitch angles. Data are displayed as a function of universal time (UT), magnetic local time (MLT), geomagnetic latitude (mlat), and L shell, respectively.

Figure 8 shows some typical half-orbit data of the CSES under a quiet space weather condition recorded by orbit No. 0030970 on 24 August 2018 (denoted by the vertical line Figure 1). The ELF/VLF waves are presented by the PSD values of the magnetic and electric fields (Figure 8a,b), respectively. The overlapping dashed curves represent the local proton cyclotron frequency f_{cp}, which was computed by the total magnetic field values provided by the HPM onboard the CSES.

From the energy spectrum of energetic electrons at 0.1 MeV < E < 3 MeV (Figure 8c), the outer/inner radiation belt is clearly visible (denoted by black arrows). In addition, we can see from Figure 8 that even under quiet space weather conditions at geomagnetic latitudes between ~40° and ~70°, there was a much higher level of ELF/VLF wave activity and energetic electron fluxes than the lower latitudes. Since the CSES is switched off at latitudes over ~65°, the region above 70° cannot be observed by it. Note that the electric field enhancement over the equatorial area (denoted by black rectangle in Figures 8b and 9b) was due to artificial interferences from the EFD.

The variation of energy spectrum with respect to the local pitch angle distributions was further examined, as shown in Figure 8d, by using data from the nine silicon-slice units of HEPP-L onboard the CSES. Distinctly, at high latitudes over 50°, where ELF hiss waves appeared below 900 Hz, the energetic fluxes overwhelmingly got distributed at pitch angles from ~65° to 120°. The region within geomagnetic latitudes ±30° saw that the energetic electron fluxes dramatically declined, but the wave activities simultaneously enhanced at a frequency higher than f_{cp}.

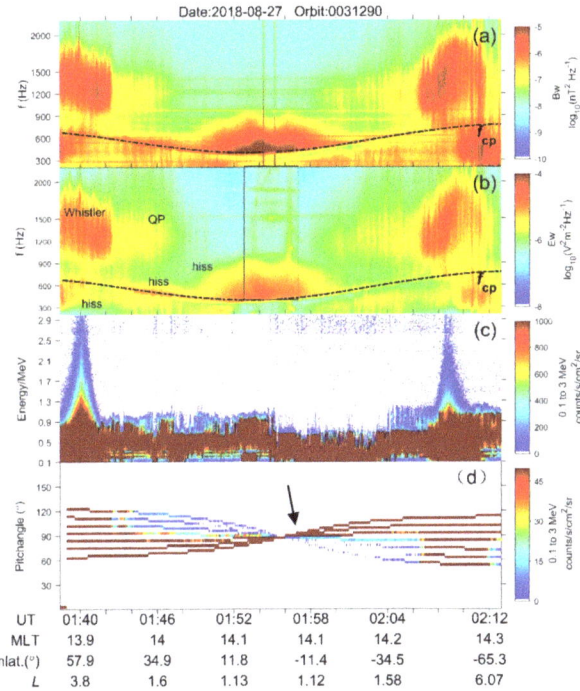

Figure 9. Same as Figure 8 but under disturbed space weather conditions in the nightside ionosphere recorded by orbit No. 0031290 on 27 August 2018.

Figure 9 shows data from another half orbit that were recorded by orbit No. 0030970 during geomagnetic storm time on 27 August 2018 (recovery phase, denoted by the vertical solid line in Figure 1). Compared to the quiet condition in Figure 8, the CSES witnessed much stronger ELF/VLF activities and energetic electron fluxes enhancement under the active storm condition. Three typical electromagnetic waves could especially be identified, as follows:

(1) The ionospheric hiss waves that mainly appeared at frequencies around/below f_{cp} at geomagnetic latitudes from 20° (relatively weak) to 65° (predominantly) or at frequencies above f_{cp} to 800 Hz over the equatorial area with 20°, showing structure-less wave spectral property [22]; (2) the non-structured VLF waves at frequencies from ~1000 to 1800 Hz at high latitudes over 55°; and (3) the structured quasi-periodic structures (QP waves) at latitudes from ~35° to 50° [20].

For energetic electron fluxes, the CSES saw a direct enhancement of $E < 0.9$ MeV along the entire orbit trace, as well as increases of particles 0.1 MeV $< E < 3$ MeV at high latitudes around ~55° to 60°. It has to be noted that there were three silicon units that got saturated (see the arrow in Figure 9d); such saturation phenomena only occur under geomagnetically disturbed conditions, and data from the saturated silicon units were eliminated in the statistical analysis from Figures 6 and 7.

In contrast to the quiet condition presented by Figure 8c, the slot region in Figure 9c is invisible; such disappearance can be ascribed to large scale precipitation extending to very low L shell region (L~2). In other words, the slot region was refilled by a large quantity of energetic particles that were probably accelerated or scattered through the ELF/VLF wave–particle interaction process in the outer radiation belt. The slot region refilling phenomena during this storm time was also investigated by Zhang et al. [26] based on a conjugated observations between the CSES and RBSP (Radiation Belt Storm Probes), and they found that the ELF/VLF whistler-mode chorus waves in the radiation belt could efficiently accelerate and diffuse the relativistic electrons at the extremely low L shell area in the ionosphere.

We further computed equatorial pitch angles by using the field-line tracing method based on the IGRF (International Geomagnetic Reference Field) model and the total magnetic field intensity recorded by the HPM onboard the CSES (see details in the work of Zhima et al. [20]). Figures 10a and 11a show the observed local pitch angles, Figures 10b and 11b the computed equatorial pitch angles for the half orbits of Figures 8 and 9. It can be seen that for geomagnetic latitudes higher than ~40°, the computed equatorial pitch angles for the local observed particles were overwhelmingly lower than the equatorial loss cone (denoted by the thick black dashed lines in Figures 10b and 11b), indicating these particles in the high latitude ionosphere as basically lost. However, at lower geomagnetic latitudes roughly within ±40°, we can see that certain particles were larger than the equatorial loss cone, indicating that they were most likely trapped instead of precipitation into this orbit space. It can be said that the CSES provided a good coverage of the energetic particle precipitations in the high latitude ionosphere.

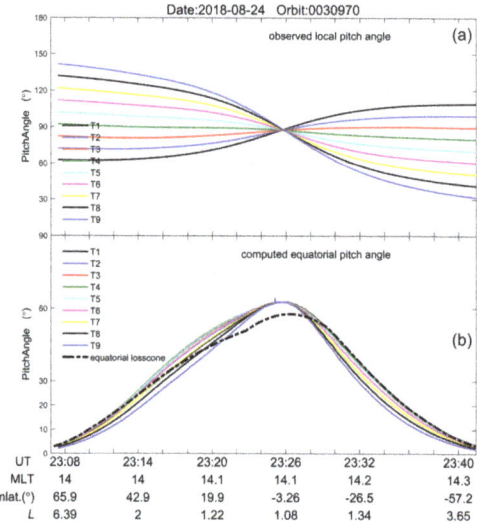

Figure 10. The comparison between the observed local pitch angels (**a**) and the computed equatorial pitch angles (**b**) for orbit No. 0030970 on 24 August 2018. The thick dashed line in (**b**) displays the equatorial loss cone (see text). T1–T9 with different colors in the legend denote the results from the nine silicon-slice units of HEPP-L onboard the CSES.

After investigating a large amount of half orbit data (with every day of 62 half orbits) from 20 August to 4 September 2018, some basic features of energetic electron fluxes could be recovered: (1) under the quiet condition, the CSES could well depict the outer/inner radiation belt and the slot region (as Figure 8) well, whereas under disturbed conditions, such a region boundary was invisible due to large scale precipitation extending to very low L shells; (2) the regions poleward from geomagnetic latitudes 50° corresponded to the highest electron precipitation without any distinction between solar quiet or active conditions, and the regions below geomagnetic latitudes 30° generally got precipitations during the storm time and occasionally at quiet time; and (3) the ELF/VLF waves recorded by the CSES mainly included structure-less VLF whistler waves, structured quasi-periodic emissions, and structure-less ELF hiss waves. On the contrary, no whistler-mode chorus waves were found during this storm. Specifically, Figure 9 shows an example of the existence of whistler-mode waves with both non-structured (at the geomagnetic latitudes over 55°) and structured QP waves (at latitudes of 30° to 40°) at frequencies from 900 to 1200 Hz, as well as ELF hiss waves below the local proton cyclotron frequency f_{cp}.

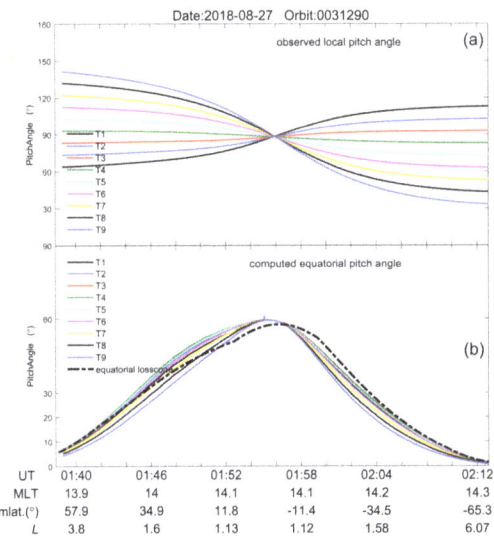

Figure 11. Same as Figure 10 but for orbit No. 0031290 on 27 August 2018.

Thanks to the availability of waveform data below 2.5 kHz along the whole orbit trajectory, we could conduct a wave vector analysis to discuss the propagation features for these typical ELF/VLF whistler-mode emissions that occurred during the storm time. The wave propagation parameters for the ELF/VLF waves are shown in Figure 12, and they were computed by the singular value decomposition method [46] under the geomagnetic field aligned coordinate system (FAC) of the CSES (see details in Section 3.2 of the work of Zhima et al. [20]).

Figure 12 shows the wave vector analysis results for the ELF/VLF waves at northern high geomagnetic latitude regions (corresponding to the right part of Figure 9). The ellipticity value (Figure 12c), which is the ratio of the axes of polarization ellipse (+1 means right-hand circular polarization, −1 means left-hand circular polarization, and 0 means linear polarization), indicated that waves (900 Hz < f < 2000 Hz) and hiss waves ($f < f_{cp}$) were of the right-handed polarized whistler-mode and their wave normal angles (θ_k) (Figure 12d) generally varied in the range of around 40° to 65°. This suggested that waves obliquely propagated to the background magnetic field. Figure 12e shows the azimuthal angles (φ_k) (±180°: decreasing L shell direction; 0°: increasing L shell direction), which present values near 180°, suggesting that the observed ELF/VLF waves propagated towards the Earth direction (that is, in the decreasing L shell direction). The planarity of waves (Figure 12f), which represents the wave propagation mode, exhibited a value mostly equal to +1, meaning that the observed waves were coming towards the spacecraft as plane waves. The Poynting fluxes ($E \times H$) were computed by the six components of the electromagnetic waves [20]. It can be seen from Figure 12 that Poynting fluxes (S) mainly dominated in the perpendicular direction to the background magnetic field instead of the parallel directions.

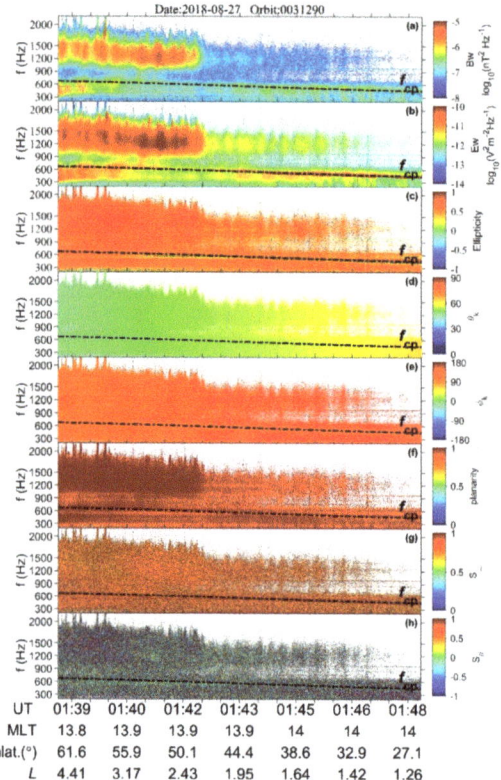

Figure 12. The wave propagation parameters computed by the singular value decomposition (SVD) method for the ELF/VLF waves in the high-latitude ionosphere during the geomagnetic storm recorded by orbit No. 0031290 on 27 August 2018. From top to bottom: the sum of the PSD values of the three components of the magnetic field (**a**) and the electric field (**b**); the ellipticity (**c**); the wave normal angle (**d**) and the azimuthal angle (**e**) for wave vector k; and the planarity (**f**) and the perpendicular and parallel component of the Poynting vector ($S_\perp$, $S_{||}$) (**g**,**h**).

According to Figure 12, the ELF/VLF whistler-mode waves obliquely propagated from the outer radiation belt towards the satellite along the Earth direction. Such peculiarity agreed with previous studies [17,29] that highlighted that in the high latitude ionosphere, these particles most likely propagate from the outer radiation belt and precipitate into the ionosphere (or into the atmosphere) as a consequence of their interactions with ELF/VLF waves. Zhima et al. [9] and Fu et al. [47] interpreted the generation of ELF/VLF waves in the ionosphere as mostly likely being due to the strong temperature anisotropy after solar wind energetic particle injections during storm time, which provides free energy for wave excitation to amplify the ELF/VLF waves in the upper ionosphere.

5. Conclusions

This study reports the temporal and spatial distributions of ELF/VLF wave activities and energetic particle enhancement in the mid-high latitude ionosphere during storm time based on observations from the SCM, EFD, and HEPP-L payloads onboard the CSES, which has been operating in low Earth orbit at an altitude of 507 km from February 2018 to present.

The 26 August 2018 intense storm resulted from an ICME event from the Sun, leading to strong ELF/VLF emissions and energetic electron precipitations in the upper ionosphere.

The superposed epoch analysis for ELF/VLF waves and particles indicated that before the ICME hit the magnetopause, the wave activities and particle precipitation were relatively weak in the ionosphere, and the climax of wave excitations mainly appeared during the near-end main phase and the early recovery phase. Regarding frequency, the waves at frequencies below 6 kHz mainly occurred at the early stage of the main phase; a broad frequency band wave from f_{cp} to 20 kHz was remarkably excited from the near-end main phase to the early recovery phase. During the long period of the recovery phase, the wave enhancements mainly occurred below 6 kHz. The wave activity in the nightside ionosphere showed a good correlation with the evolution of the geomagnetic storm, but the amplitude of such was weaker than that on the dayside its spatial scale was also narrowly distributed.

The energetic precipitating fluxes increased with geomagnetic activity and reached their maximum during the early recovery phase, as primarily observed outside of the plasmapause. The energetic electrons at an energy level below 1.5 MeV got strong enhancements during the entire storm time on both the day and night side (relatively stronger than the ones on the dayside), and they also appeared both inside and outside the plasmapause; for particles $E > 1.5$ MeV, the fluxes mainly got enhancement from outside the plasmapause.

An investigation into a large amount of half orbit data in the whole time period showed that under the quiet condition, the CSES depicted the outer/inner radiation belt in the slot region well, whereas under disturbed conditions, such regions were invisible and dominated by precipitations of $E < 1$ MeV all along the CSES orbit space. The regions poleward from the geomagnetic latitude of 50° corresponded to the highest electron precipitation regardless of the quiet or active conditions; the regions equatorward below 30° were usually enhanced during the storm time and occasionally during the quiet time.

The ELF/VLF waves recorded by the CSES during the storm time mainly included structure-less VLF waves, structured ELF/VLF quasi-periodic emissions, and non-structured ELF hiss waves. A wave vector analysis indicated that the ELF/VLF waves were right-handed polarized whistler-mode waves, and they obliquely propagated from some area higher than the satellite (most likely from the outer radiation belt) to the Earth direction.

Author Contributions: Conceptualization, Z.Z. (Zeren Zhima); data curation, Y.H., W.C., M.P., A.P., Z.Z. (Zhenxia Zhang), Q.W., J.H., S.Z., Y.Y., D.Y., X.S. (Xiaoying Sun), Q.T., N.Z., and F.G.; formal analysis, Z.Z. (Zeren Zhima), W.C., M.P., A.P., and Z.Z. (Zhenxia Zhang); funding acquisition, Z.Z. (Zeren Zhima); investigation, Z.Z. (Zeren Zhima), Y.H., W.C., M.P., A.P., Z.Z. (Zhenxia Zhang), Q.W., J.H., S.Z., Y.Y., D.Y., and X.S. (Xiaoying Sun); methodology, Z.Z. (Zeren Zhima), A.P., Z.Z. (Zhenxia Zhang), Q.W., and Y.Y.; project administration, Z.Z. (Zeren Zhima); resources, Z.Z. (Zeren Zhima), X.S. (Xuhui Shen), Q.T., N.Z., and F.G.; software, Z.Z. (Zeren Zhima) and Y.H.; supervision, Z.Z. (Zeren Zhima); validation, Z.Z. (Zeren Zhima), W.C., M.P., A.P., and Z.Z. (Zhenxia Zhang); visualization, Z.Z. (Zeren Zhima); writing—original draft, Z.Z. (Zeren Zhima); writing—review and editing, Z.Z. (Zeren Zhima). All authors have read and agreed to the published version of the manuscript.

Funding: This work is supported by the NSFC Grant 41874174/41574139, National Key R&D Program of China (Grant No.2018YFC1503501), the APSCO Earthquake Research Project Phase II and ISSI-BJ project. M. Piersanti and Alexandra Parmentier thank the Italian Space Agency (ASI) for the financial support under the contract ASI "LIMADOU scienza" no 2016-16-H0.

Data Availability Statement: The data of HPM, EFD, SCM, HEPP-L on board CSES can be download from the website (http://www.leos.ac.cn/, accessed on 1 December 2020). The solar wind parameters and geomagnetic Kp and Dst indices are provided by the OMINI database (https://cdaweb.sci.gsfc.nasa.gov/index.html/, accessed on 1 December 2020).

Conflicts of Interest: The authors declare no conflict of interest.

References

1. Pulinets, S.; Ouzounov, D. Lithosphere–Atmosphere–Ionosphere Coupling (LAIC) model—An unified concept for earthquake precursors validation. *J. Asian Earth Sci.* **2011**, *41*, 371–382. [CrossRef]
2. Zhao, S.; Zhou, C.; Shen, X.; Zhima, Z. Investigation of VLF Transmitter Signals in the Ionosphere by ZH-1 Observations and Full-Wave Simulation. *J. Geophys. Res. Space Phys.* **2019**, *124*, 4697–4709. [CrossRef]

3. Parrot, M. DEMETER observations of manmade waves that propagate in the ionosphere. *Comptes Rendus Phys.* **2018**, *19*, 26–35. [CrossRef]
4. Piersanti, M.; Alberti, T.; Bemporad, A.; Berrilli, F.; Bruno, R.; Capparelli, V.; Carbone, V.; Cesaroni, C.; Consolini, G.; Cristaldi, A.; et al. Comprehensive analysis of the geoeffective solar event of 21 June 2015: Effects on the magnetosphere, plasmasphere, and ionosphere systems. *Solar Phys.* **2017**, *292*, 1–56. [CrossRef]
5. Yang, Y.-Y.; Zhima, Z.-R.; Shen, X.-H.; Chu, W.; Huang, J.-P.; Wang, Q.; Yan, R.; Xu, S.; Lu, H.-X.; Liu, D.-P. The First Intense Geomagnetic Storm Event Recorded by the China Seismo-Electromagnetic Satellite. *Space Weather* **2020**, *18*, e2019SW002243. [CrossRef]
6. Piersanti, M.; Materassi, M.; Battiston, R.; Carbone, V.; Cicone, A.; D'Angelo, G.; Diego, P.; Ubertini, P. Magnetospheric–Ionospheric–Lithospheric Coupling Model. 1: Observations during the 5 August 2018 Bayan Earthquake. *Remote. Sens.* **2020**, *12*, 3299. [CrossRef]
7. Piersanti, M.; De Michelis, P.; Del Moro, D.; Tozzi, R.; Pezzopane, M.; Consolini, G.; Marcucci, M.F.; Laurenza, M.; Di Matteo, S.; Pignalberi, A.; et al. From the Sun to Earth: Effects of the 25 August 2018 geomagnetic storm. *Ann. Geophys.* **2020**, *38*, 703–724. [CrossRef]
8. Zhima, Z.; Hu, Y.; Piersanti, M.; Shen, X.; De Santis, A.; Yan, R.; Yang, Y.; Zhao, S.; Zhang, Z.; Wang, Q.; et al. The Seismic Electromagnetic Emissions during the 2010 Mw 7.8 Northern Sumatra Earthquake Revealed by DEMETER Satellite. *Front. Earth Sci.* **2020**, *8*, 459. [CrossRef]
9. Zhima, Z.; Cao, J.; Liu, W.; Fu, H.; Wang, T.; Zhang, X.; Shen, X. Storm time evolution of ELF/VLF waves observed by DEMETER satellite. *J. Geophys. Res. Space Phys.* **2014**, *119*, 2612–2622. [CrossRef]
10. Zhima, Z.; Cao, J.; Liu, W.; Fu, H.; Yang, J.; Zhang, X.; Shen, X. DEMETER observations of high-latitude chorus waves penetrating the plasmasphere during a geomagnetic storm. *Geophys. Res. Lett.* **2013**, *40*, 5827–5832. [CrossRef]
11. Parrot, M.; Berthelier, J.; Lebreton, J.; Sauvaud, J.; Santolik, O.; Blecki, J. Examples of unusual ionospheric observations made by the DEMETER satellite over seismic regions. *Phys. Chem. Earth Parts A/B/C* **2006**, *31*, 486–495. [CrossRef]
12. Santolík, O.; Chum, J.; Parrot, M.; Gurnett, D.A.; Pickett, J.S.; Cornilleau-Wehrlin, N. Propagation of whistler mode chorus to low altitudes: Spacecraft observations of structured ELF hiss. *J. Geophys. Res. Space Phys.* **2006**, *111*, 10208. [CrossRef]
13. Němec, F.; Bezděková, B.; Manninen, J.; Parrot, M.; Santolík, O.; Hayosh, M.; Turunen, T. Conjugate observations of a remarkable quasiperiodic event by the low-altitude DEMETER spacecraft and ground-based instruments. *J. Geophys. Res. Space Phys.* **2016**, *121*, 8790–8803. [CrossRef]
14. Summers, D.; Ni, B.; Meredith, N.P.; Thorne, R.M.; Moldwin, M.B.; Anderson, R.R. Electron scattering by whistler-mode ELF hiss in plasmaspheric plumes. *J. Geophys. Res. Space Phys.* **2008**, *113*, 04219. [CrossRef]
15. Hayosh, M.; Pasmanik, D.L.; Demekhov, A.G.; Santolik, O.; Parrot, M.; Titova, E.E. Simultaneous observations of quasi-periodic ELF/VLF wave emissions and electron precipitation by DEMETER satellite: A case study. *J. Geophys. Res. Space Phys.* **2013**, *118*, 4523–4533. [CrossRef]
16. Parrot, M.; Santolík, O.; Němec, F. Chorus and chorus-like emissions seen by the ionospheric satellite DEMETER. *J. Geophys. Res. Space Phys.* **2016**, *121*, 3781–3792. [CrossRef]
17. Horne, R.B.; Shprits, Y.Y.; Meredith, N.P.; Glauert, S.A.; Smith, A.J.; Kanekal, S.G.; Baker, D.N.; Engebretson, M.J.; Posch, J.L.; Spasojevic, M.; et al. Wave acceleration of electrons in the Van Allen radiation belts. *Nat. Cell Biol.* **2005**, *437*, 227–230. [CrossRef]
18. Benck, S.; Cyamukungu, M.; Cabrera, J. Study of correlations between waves and particle fluxes measured on board the DEMETER satellite. *Adv. Space Res.* **2008**, *42*, 1538–1549. [CrossRef]
19. Cao, J.B.; Liu, Z.X.; Yan, C.X.; Cai, C.L.; Li, Z.Y.; Zhu, G.W.; Wang, S.R.; Zhao, H.; Liang, J.B.; Ren, Q.Y.; et al. First results of Chinese particle instruments in the Double Star Program. *Ann. Geophys.* **2005**, *23*, 2775–2784. [CrossRef]
20. Zhima, Z.; Huang, J.; Shen, X.; Xia, Z.; Chen, L.; Piersanti, M.; Yang, Y.; Wang, Q.; Zeng, L.; Lei, J.; et al. Simultaneous Observations of ELF/VLF Rising-Tone Quasiperiodic Waves and Energetic Electron Precipitations in the High-Latitude Upper Ionosphere. *J. Geophys. Res. Space Phys.* **2020**, *125*, e2019JA027574. [CrossRef]
21. Chen, L.; Santolík, O.; Hajoš, M.; Zheng, L.; Zhima, Z.; Heelis, R.; Hanzelka, M.; Horne, R.B.; Parrot, M. Source of the low-altitude hiss in the ionosphere. *Geophys. Res. Lett.* **2017**, *44*, 2060–2069. [CrossRef]
22. Zhima, Z.; Chen, L.; Xiong, Y.; Cao, J.; Fu, H. On the Origin of Ionospheric Hiss: A Conjugate Observation. *J. Geophys. Res. Space Phys.* **2017**, *122*, 11784–11793. [CrossRef]
23. Xia, Z.; Chen, L.; Zhima, Z.; Santolík, O.; Horne, R.B.; Parrot, M. Statistical Characteristics of Ionospheric Hiss Waves. *Geophys. Res. Lett.* **2019**, *46*, 7147–7156. [CrossRef]
24. Santolik, O.; Parrot, M.; Inan, U.S.; Burešová, D.; Gurnett, D.A.; Chum, J. Propagation of unducted whistlers from their source lightning: A case study. *J. Geophys. Res. Space Phys.* **2009**, *114*. [CrossRef]
25. Rodger, C.J.; Clilverd, M.A.; Thomson, N.R.; Gamble, R.J.; Seppälä, A.; Turunen, E.; Meredith, N.P.; Parrot, M.; Sauvaud, J.-A.; Berthelier, J.-J. Radiation belt electron precipitation into the atmosphere: Recovery from a geomagnetic storm. *J. Geophys. Res. Space Phys.* **2007**, *112*. [CrossRef]
26. Zhang, Z.; Chen, L.; Liu, S.; Xiong, Y.; Li, X.; Wang, Y.; Chu, W.; Zeren, Z.; Shen, X. Chorus Acceleration of Relativistic Electrons in Extremely Low L-Shell During Geomagnetic Storm of August. *Geophys. Res. Lett.* **2020**, *47*, e2019GL086226. [CrossRef]
27. Thorne, R.M.; Smith, E.J.; Burton, R.K.; Holzer, R.E. Plasmaspheric hiss. *J. Geophys. Res. Space Phys.* **1973**, *78*, 1581–1596. [CrossRef]

28. Chen, L.; Bortnik, J.; Li, W.; Thorne, R.M.; Horne, R.B. Modeling the properties of plasmaspheric hiss: 1. Dependence on chorus wave emission. *J. Geophys. Res. Space Phys.* **2012**, *117*. [CrossRef]
29. Summers, D.; Ni, B.; Meredith, N.P. Timescales for radiation belt electron acceleration and loss due to resonant wave-particle interactions: 2. Evaluation for VLF chorus, ELF hiss, and electromagnetic ion cyclotron waves. *J. Geophys. Res. Space Phys.* **2007**, *112*, 04207. [CrossRef]
30. Parrot, M.; Benoist, D.; Berthelier, J.; Błęcki, J.; Chapuis, Y.; Colin, F.; Elie, F.; Fergeau, P.; Lagoutte, D.; Lefeuvre, F.; et al. The magnetic field experiment IMSC and its data processing onboard DEMETER: Scientific objectives, description and first results. *Planet. Space Sci.* **2006**, *54*, 441–455. [CrossRef]
31. Zhao, S.; Shen, X.; Zhou, C.; Liao, L.; Zhima, Z.; Wang, F. The influence of the ionospheric disturbance on the ground based VLF transmitter signal recorded by LEO satellite—Insight from full wave simulation. *Results Phys.* **2020**, *19*, 103391. [CrossRef]
32. Yan, R.; Shen, X.; Huang, J.; Wang, Q.; Chu, W.; Liu, D.; Yang, Y.; Lu, H.; Xu, S. Examples of unusual ionospheric observations by the CSES prior to earthquakes. *Earth Planet. Phys.* **2018**, *2*, 515–526. [CrossRef]
33. Shen, X.; Zhang, X.; Yuan, S.; Wang, L.; Cao, J.; Huang, J.; Zhu, X.; Piergiorgio, P.; Dai, J. The state-of-the-art of the China Seismo-Electromagnetic Satellite mission. *Sci. China Ser. E: Technol. Sci.* **2018**, *61*, 634–642. [CrossRef]
34. Shen, X.; Zong, Q.-G.; Zhang, X. Introduction to special section on the China Seismo-Electromagnetic Satellite and initial results. *Earth Planet. Phys.* **2018**, *2*, 439–443. [CrossRef]
35. Zhou, B.; Yang, Y.; Zhang, Y.; Gou, X.; Cheng, B.; Wang, J.; Li, L. Magnetic field data processing methods of the China Seismo-Electromagnetic Satellite. *Earth Planet. Phys.* **2018**, *2*, 455–461. [CrossRef]
36. Cheng, B.; Zhou, B.; Magnes, W.; Lammegger, R.; Pollinger, A. High precision magnetometer for geomagnetic exploration onboard of the China Seismo-Electromagnetic Satellite. *Sci. China Ser. E: Technol. Sci.* **2018**, *61*, 659–668. [CrossRef]
37. Pollinger, A.; Lammegger, R.; Magnes, W.; Hagen, C.; Ellmeier, M.; Jernej, I.; Leichtfried, M.; Kürbisch, C.; Maierhofer, R.; Wallner, R.; et al. Coupled dark state magnetometer for the China Seismo-Electromagnetic Satellite. *Meas. Sci. Technol.* **2018**, *29*, 095103. [CrossRef]
38. Cao, J.; Zeng, L.; Zhan, F.; Wang, Z.; Wang, Y.; Chen, Y.; Meng, Q.; Ji, Z.; Wang, P.; Liu, Z.; et al. The electromagnetic wave experiment for CSES mission: Search coil magnetometer. *Sci. China Ser. E: Technol. Sci.* **2018**, *61*, 653–658. [CrossRef]
39. Wang, Q.; Huang, J.; Zhang, X.; Shen, X.; Yuan, S.; Zeng, L.; Cao, J. China Seismo-Electromagnetic Satellite search coil magnetometer data and initial results. *Earth Planet. Phys.* **2018**, *2*, 1–7. [CrossRef]
40. Huang, J.; Lei, J.; Li, S.; Zeren, Z.; Li, C.; Zhu, X.; Yu, W. The Electric Field Detector (EFD) onboard the ZH-1 satellite and first observational results. *Earth Planet. Phys.* **2018**, *2*, 469–478. [CrossRef]
41. Chu, W.; Huang, J.; Shen, X.; Wang, P.; Li, X.; An, Z.; Xu, Y.; Liang, X. Preliminary results of the High Energetic Particle Package on-board the China Seismo-Electromagnetic Satellite. *Earth Planet. Phys.* **2018**, *2*, 1–10. [CrossRef]
42. Li, X.Q.; Xu, Y.B.; An, Z.H.; Liang, X.H.; Wang, P.; Zhao, X.Y.; Wang, H.Y.; Lu, H.; Ma, Y.Q.; Shen, X.H.; et al. The high-energy particle package onboard CSES. *Radiat. Detect. Technol. Methods* **2019**, *3*, 22. [CrossRef]
43. Piersanti, M.; Pezzopane, M.; Zhima, Z.; Diego, P.; Xiong, C.; Tozzi, R.; Pignalberi, A.; D'Angelo, G.; Battiston, R.; Huang, J.; et al. Can an impulsive variation of the solar wind plasma pressure trigger a plasma bubble? A case study based on CSES, Swarm and THEMIS data. *Adv. Space Res.* **2021**, *67*, 35–45. [CrossRef]
44. Younas, W.; Amory-Mazaudier, C.; Khan, M.; Fleury, R. Ionospheric and Magnetic Signatures of a Space Weather Event on 25–29 August 2018: CME and HSSWs. *J. Geophys. Res. Space Phys.* **2020**, *125*, e2020JA027981. [CrossRef]
45. O'Brien, T.P.; Moldwin, M.B. Empirical plasmapause models from magnetic indices. *Geophys. Res. Lett.* **2003**, *30*, 1152. [CrossRef]
46. Santolik, O.; Parrot, M.; Lefeuvre, F. Singular value decomposition methods for wave propagation analysis. *Radio Sci.* **2003**, *38*. [CrossRef]
47. Fu, H.S.; Cao, J.B.; Mozer, F.S.; Lu, H.Y.; Yang, B. Chorus intensification in response to interplanetary shock. *J. Geophys. Res. Space Phys.* **2012**, *117*, 01203. [CrossRef]

Article

Trapped Proton Fluxes Estimation Inside the South Atlantic Anomaly Using the NASA AE9/AP9/SPM Radiation Models along the China Seismo-Electromagnetic Satellite Orbit

Matteo Martucci [1,2,*], Roberta Sparvoli [1,2], Simona Bartocci [1], Roberto Battiston [3,4], William Jerome Burger [4,5], Donatella Campana [6], Luca Carfora [1,2], Guido Castellini [7], Livio Conti [1,8], Andrea Contin [9,10], Cinzia De Donato [1], Cristian De Santis [1], Francesco Maria Follega [3,4], Roberto Iuppa [3,4], Ignazio Lazzizzera [3,4], Nadir Marcelli [1,2], Giuseppe Masciantonio [1], Matteo Mergé [1,†], Alberto Oliva [10], Giuseppe Osteria [6], Francesco Palma [1,†], Federico Palmonari [9,10], Beatrice Panico [6], Alexandra Parmentier [1], Francesco Perfetto [6], Piergiorgio Picozza [1,2], Mirko Piersanti [11], Michele Pozzato [10], Ester Ricci [3,4], Marco Ricci [12], Sergio Bruno Ricciarini [7], Zouleikha Sahnoun [10], Valentina Scotti [6,13], Alessandro Sotgiu [1], Vincenzo Vitale [1], Simona Zoffoli [14] and Paolo Zuccon [3,4]

1. INFN-Sezione di Roma "Tor Vergata", V. della Ricerca Scientifica 1, I-00133 Rome, Italy; roberta.sparvoli@roma2.infn.it (R.S.); simona.bartocci@roma2.infn.it (S.B.); luca.carfora@roma2.infn.it (L.C.); livio.conti@uninettunouniversity.net (L.C.); cinzia.dedonato@roma2.infn.it (C.D.D.); cristian.desantis@roma2.infn.it (C.D.S.); nadir.marcelli@roma2.infn.it (N.M.); giuseppe.masciantonio@roma2.infn.it (G.M.); matteo.merge@roma2.infn.it (M.M.); francesco.palma@roma2.infn.it (F.P.); alexandra.parmentier@roma2.infn.it (A.P.); piergiorgio.picozza@roma2.infn.it (P.P.); alessandro.sotgiu@roma2.infn.it (A.S.); vincenzo.vitale@roma2.infn.it (V.V.)
2. Department of Physics, University of Rome "Tor Vergata", V. della Ricerca Scientifica 1, I-00133 Rome, Italy
3. Department of Physics, University of Trento, V. Sommarive 14, I-38123 Trento, Italy; roberto.battiston@unitn.it (R.B.); francesco.follega@unitn.it (F.M.F.); roberto.iuppa@unitn.it (R.I.); ignazio.lazzizzera@unitn.it (I.L.); ester.ricci@unitn.it (E.R.); paolo.zuccon@unitn.it (P.Z.)
4. INFN-TIFPA, V. Sommarive 14, I-38123 Trento, Italy; william.burger@tifpa.infn.it
5. Centro Fermi, V. Panisperna 89a, I-00184 Rome, Italy
6. INFN-Sezione di Napoli, V. Cintia, I-80126 Naples, Italy; donatella.campana@na.infn.it (D.C.); giuseppe.osteria@na.infn.it (G.O.); beatrice.panico@na.infn.it (B.P.); francesco.perfetto@na.infn.it (F.P.); valentina.scotti@na.infn.it (V.S.)
7. IFAC-CNR, V. Madonna del Piano 10, I-50019 Florence, Italy; g.castellini@ifac.cnr.it (G.C.); s.ricciarini@ifac.cnr.it (S.B.R.)
8. Department of Engineering, Uninettuno University, C.so V. Emanuele II 39, I-00186 Rome, Italy
9. Department of Physics, University of Bologna, V.le C. Berti Pichat 6/2, I-40127 Bologna, Italy; Andrea.Contin@bo.infn.it (A.C.); federico.palmonari@bo.infn.it (F.P.)
10. INFN-Sezione di Bologna, V.le C. Berti Pichat 6/2, I-40127 Bologna, Italy; Alberto.Oliva@bo.infn.it (A.O.); michele.pozzato@bo.infn.it (M.P.); zouleikha.sahnoun@bo.infn.it (Z.S.)
11. INAF-IAPS, V. Fosso del Cavaliere 100, I-00133 Rome, Italy; mirko.piersanti@roma2.infn.it
12. INFN-LNF, V. E. Fermi 54, I-00044 Rome, Italy; marco.ricci@lnf.infn.it
13. Department of Physics, University of Naples "Federico II", V. Cintia 21, I-80126 Naples, Italy
14. Italian Space Agency, V. del Politecnico, I-00133 Rome, Italy; simona.zoffoli@asi.it
* Correspondence: matteo.martucci@roma2.infn.it
† Also at ASI Space Science Data Center (SSDC), V. del Politecnico, I-00133 Rome, Italy.

Citation: Martucci, M.; Sparvoli, R.; Bartocci, S.; Battiston, R.; Burger, W.J.; Campana, D.; Carfora, L.; Castellini, G.; Conti, L.; Contin, A.; et al. Trapped Proton Fluxes Estimation Inside the South Atlantic Anomaly Using the NASA AE9/AP9/SPM Radiation Models along the China Seismo-Electromagnetic Satellite Orbit. *Appl. Sci.* **2021**, *11*, 3465. https://doi.org/10.3390/app11083465

Received: 9 March 2021
Accepted: 8 April 2021
Published: 13 April 2021

Publisher's Note: MDPI stays neutral with regard to jurisdictional claims in published maps and institutional affiliations.

Copyright: © 2021 by the authors. Licensee MDPI, Basel, Switzerland. This article is an open access article distributed under the terms and conditions of the Creative Commons Attribution (CC BY) license (https://creativecommons.org/licenses/by/4.0/).

Abstract: The radiation belts in the Earth's magnetosphere pose a hazard to satellite systems and spacecraft missions (both manned and unmanned), heavily affecting payload design and resources, thus resulting in an impact on the overall mission performance and final costs. The NASA AE9/AP9/SPM radiation models for energetic electrons, protons, and plasma provide useful information on the near-Earth environment, but they are still incomplete as to some features and, for some energy ranges, their predictions are not based on a statistically sufficient sample of direct measurements. Therefore, it is of the upmost importance to provide new data and direct measurements to improve their output. In this work, the AP9 model is applied to the China Seismo-Electromagnetic Satellite (CSES-01) orbit to estimate the flux of energetic protons over the South Atlantic Anomaly during a short testing period of one day, 1 January 2021. Moreover, a preliminary comparison with proton data obtained from the High-Energy Particle Detector (HEPD) on board CSES-01 is carried out. This estimation will serve as the starting ground for a forthcoming complete data analysis, enabling extensive testing and validation of current theoretical and empirical models.

Keywords: trapped particles; South Atlantic Anomaly; AE9/AP9/SPM models; radiation belts

1. Introduction

The radiation belts, also known as Van Allen belts, are regions of the Earth's magnetosphere where energetic charged particles are subject to long-term magnetic trapping. The outer belt is mostly populated by electrons with hundreds of keV to MeV energies, while the inner belt mostly consists of an intense radiation of energetic protons (from MeV up to a few GeV), electrons/positrons (up to ∼8 MeV), and a minor component of ions [1,2]. Proton populations with energies above a few tens of MeV originate from the β-decay of free neutrons produced in the interaction between galactic cosmic-rays and the Earth's atmosphere in a mechanism called *Cosmic Ray Albedo Neutron Decay* (CRAND) [3,4]. Since the discovery of the Van Allen radiation belts, after the launch of the first Explorer satellites in 1958 and the Pioneer in 1959 [5,6], the scientific community has been considerably involved in modeling this space radiation environment. All these efforts were mostly aimed to meet the practical need of better understanding the significant radiation hazard to spacecraft and human crews. Several studies reported a direct association between the dynamic radiation environment and system or subsystem performances [7,8]. To address and solve these problems, more accurate, comprehensive, and up-to-date space radiation environment models have been developed by the National Reconnaissance Office (NRO) and the Air Force Research Laboratory (AFRL); the new AE9/AP9/SPM set of models for high-energy electrons, protons, and space plasma, respectively, is derived from 45 datasets obtained from sensors on board various satellites. These datasets have been processed to create maps of the particle fluxes along with estimates of uncertainties from both imperfect measurements and space weather variability [9]. A detailed comparison between the older AE8/AP8 and the newer AE9/AP9 models is reported in [10].

Gradual deterioration of spacecraft systems and components—and their overall performances—with accumulated dose is a fact, and various failures, due to phenomena associated with Single Event Effects (SEEs) or electrostatic discharge, are particularly common.

The first empirical models of the radiation belts, developed in the 1960s and 1970s by NASA, tried to describe and represent the radiation environment and their early versions, namely, AE8 and AP8 [11], were widely employed in spite of their limitations, especially at low altitudes [11–13]. Despite being successful in describing the radiation environment, even AE9/AP9/SPM are partly incomplete, and often their predictions are not based on a statistically sufficient sample of direct measurements [10]. For this reason, it is of key importance to test them and, above all, to provide new and reliable datasets from in-flight instruments to improve their output and accuracy. Among the scientific payloads on board the China Seismo-Electromagnetic Satellite (CSES-01), in Low-Earth Orbit since February 2018, the High-Energy Particle Detector (HEPD) has gone through an intense period of testing and calibration, and it is able to measure >3 MeV electrons and >35 MeV protons with high efficiency. With an overall expected mission duration of >5 years, and together with other similar missions planned in the coming years, measurements from HEPD could enable the testing and validation of the aforementioned models. In this work, we have used orbital information from CSES-01 and the AP9 model to estimate the flux of trapped protons over the South Atlantic Anomaly (SAA) during a short testing period, i.e., January 1, 2021, in order to assess the radiation level at CSES orbit in view of a comparison to experimental HEPD data in a forthcoming publication. A brief description of the South Atlantic Anomaly is given in Section 2, while some details on the CSES mission and the

HEPD payload are given in Section 3. The analysis is described in Section 4, results are presented in Section 5, and, finally, a brief discussion is presented in Section 6.

2. The South Atlantic Anomaly

The South Atlantic Anomaly (SAA) is one of the most well-known features of the Earth's magnetic field. It emerges as a consequence of the tilt ($\sim 10°$) between the magnetic dipole axis of the Earth and its rotational axis and of the offset (~ 500 km) between the dipole and the Earth centers. It can be considered as the response of an inverse flux path at the core–mantle boundary of the radial component of the geomagnetic field—located approximately under the South Atlantic Ocean, which generates the hemisphere asymmetry of the Earth's magnetic field [14]; this region is characterized by an extremely low intensity of the geomagnetic field, and its behavior suggests that this asymmetry could be connected to the general decrease of the dipolar field and to the significant increase of the non-dipolar field in the Southern Atlantic region [15,16]. The extent area of the SAA at the surface of the Earth has been continuously growing since instrumental intensity measurements were made available. Several studies relate this as an indicator of a possible upcoming geomagnetic transition (excursion or reversal). It is generally accepted that such transitions are anticipated by flux patches of reversed polarity, slowly appearing at low- or mid-latitude, which migrate towards the pole [17,18].

The spatial and temporal evolution of the geomagnetic field has been monitored since 1832, when Carl-Friedrich Gauss performed the first intensity measurements in this region [19]. It has been shown that the magnetic dipole strength has been continuously decreasing [20], and data from the Swarm mission [21] revealed that two different patches are present over South America and near the coast of Africa, the latter growing at a rate of -2.54×10^5 nT per century [22]. A correct modelization of the SAA is of capital importance due to the high impact it has on human health and on instrumental efficiency [23]. Furthermore, recent studies indicate that the extent of the anomaly follows a log-periodic acceleration, resembling the behavior of a system that moves toward a critical transition [24].

3. The CSES Scientific Mission

The China Seismo-Electromagnetic Satellite (CSES-01) [25] is the first of a series of multi-instrument monitoring satellites scheduled for launch in the next few years; it is designed to study the near-Earth environment, addressing variations of the electromagnetic field, plasma parameters, and particle fluxes linked to natural sources or artificial emitters. The main scientific objective of this mission—resulting from a Chinese/Italian joint effort—is to investigate possible correlations between the aforementioned perturbations and the occurrence of high-magnitude seismic events, but it is also well suited for studying a wide variety of space-weather phenomena triggered by solar–terrestrial interactions on short (i.e., geomagnetic storms, solar particle events, etc.) and long time-scales (i.e., cosmic-ray propagation, composition, etc.) [26,27]. A recent perspective [28] explained that claims based on self-organized criticality stating that at any moment any small earthquake can eventually cascade to a large event do not stand in view of the results obtained by natural time analysis [29,30]. The CSES-01 satellite, based on the Chinese 3-axis-stabilized CAST2000 platform (total mass ~ 700 kg), is flying on a sun-synchronous polar orbit at a ~ 507 km altitude, with a 97° inclination, a period of 94.6 min, and a 5-day revisiting periodicity. Nine scientific payloads are present on board CSES: two sets of particle detectors, namely, the High-Energy Particle Package (HEPP) [31] and the High-Energy Particle Detector (HEPD) [32]; a High-Precision Magnetometer (HPM) [33]; a Search-Coil Magnetometer (SCM) [34]; an Electric Field Detector (EFD) [35]; a Global Navigation Satellite System (GNSS) Occultation Receiver [36]; a Langmuir Probe (LAP) [37]; a Tre-Band Beacon transmitter (TBB) [38]; and a Plasma Analyzer Package (PAP) [39].

The High-Energy Particle Detector

The High-Energy Particle Detector (HEPD) is a light and compact (40.36 cm × 53.00 cm × 38.15 cm, total mass ~45 kg) payload designed and built by the Limadou team, the Italian branch of the CSES Collaboration. The apparatus is made up of a series of sub-detectors:

- A tracking system, including two 213.2 mm × 214.8 mm × 0.3 mm double-sided silicon microstrip planes. Each silicon plane is divided into three identical independent sections, each of which containing two silicon sensors.
- A trigger system, consisting of one EJ-200 plastic scintillator layer segmented into six paddles (20 cm × 3 cm × 0.5 cm apiece), each one read out by two Photo-Multiplier Tubes (PMTs).
- A range calorimeter composed of two sections: The upper part is a tower of 16 EJ-200 plastic scintillator planes (15 cm × 15 cm × 1 cm), each one read out by two PMTs. The lower part is a 3 × 3 matrix of Lutetium-Yttrium Oxyorthosilicate (LYSO) inorganic scintillator crystals—5 cm × 5 cm × 4 cm each; each of the nine crystals is read out by one PMT located at its bottom side; and
- an anti-coincidence (VETO) system composed of five EJ-200 plastic scintillator planes (0.5 cm thick), each one read out by two PMTs.

The instrument is optimized to detect electrons in the 3 to 100 MeV energy range and protons between 35 and 250 MeV, as well as light nuclei. In these three years of flight, after a long period of calibration and testing, HEPD has been able to measure fluxes of low-energy galactic protons with great precision [40] and to observe the effects of the geomagnetic storm of August 2018 [41]. All the capabilities assessed in these years make HEPD well suited for the analysis of low-energy electrons and protons with good angular resolution and stability over time, which is particularly useful in highly anisotropic flux conditions like the ones encountered in SAA. More technical details can be found in [32,42,43].

4. Materials and Methods

The AE9/AP9/SPM set of models (version V1.50.001-release date December 2017) was downloaded from the Virtual Distributed Laboratory (VDL) website of the Air Force Research Laboratory (https://www.vdl.afrl.af.mil/programs/ae9ap9/, accessed on 1 February 2021). Element Set (ELSET) data—including Two-Line Elements (TLE) for the CSES satellite on 1 January 2021—have been retrieved from the Space-Track website (https://www.space-track.org/, accessed on 1 February 2021) and inserted in the code to generate the ephemeris of the satellite (at a 5 s resolution). The Simplified General Perturbations (*SGP4*) (the SGP4 propagator considers secular and periodic variations due the oblateness of the Earth, gravitational effects from Sun and Moon, and orbital drag and decay.) propagator has been preferred to the default *Kepler* with *J2* perturbation effects, for cross-checking purposes. Indeed, these orbital results have been further compared to the ones obtained using 2-min broadcast information downloaded from the satellite itself, to verify the correctness of the procedure. This cross-check includes the following:

- TLEs propagation using a chain of custom programs (including various SGP4 routines) to comply with the technique employed by the AE9/AP9 models;
- IGRF-13 [20] model-based routines have been applied to the calculated trajectory to reconstruct the intensity of the magnetic field at a 1 second resolution, together with McIllwain's L parameter in dipolar approximation [44]; these will be useful benchmarks in the future comparison to HEPD data.

After these steps, two sets of geographical/geomagnetic coordinates have been obtained: one calculated by NASA routines and another extracted by Limadou external routines used for trajectory propagation and magnetic field reconstruction since launch. A comparison between such datasets has been performed to assure the best possible agreement over the chosen testing period—January 1, 2021. The resulting discrepancies are <0.08% for LATs/LONs, <0.13% for altitude, and <0.21% for magnetic field intensity. Small discrepancies are probably due to the models using the last TLE entry for orbit propagation, unlike external code picking the TLE closest in time. After ephemeris generation, an omnidirectional differential spectrum of trapped protons as a function of kinetic energy was created using AP9. This spectrum, averaged over all orbits of a single day, is shown in Figure 1; the blue arrow represents the HEPD low-energy thresholds for protons (30 MeV). Inside the inner radiation belt, trapped electron populations present a sharp threshold at ∼8 MeV, thus electrons of higher energy are virtually nonexistent. This means that in our future analysis, trapped protons will not be affected by any low-energy electrons contamination inside the SAA, consequently improving HEPD sensitivity to protons measurements.

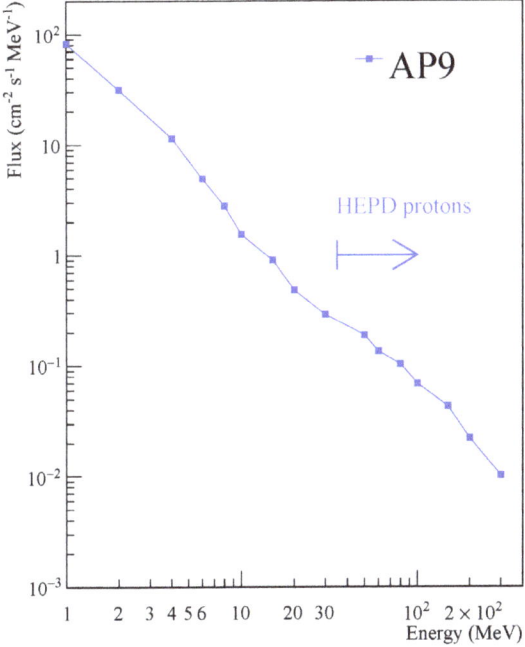

Figure 1. Omnidirectional differential spectrum of trapped protons (blue squares) as a function of kinetic energy, obtained from the AP9 model and averaged over all the orbits of the testing day—1 January 2021. HEPD low-energy threshold for protons (30 MeV) is also depicted as a blue arrow.

The AP9 model could also be very useful to help define a fiducial area on the Earth's surface (longitude vs. latitude) that may be applied to the future HEPD data analysis of trapped particles. However, the geographical extension of the inner belt (and consequently of the SAA) is largely dependent on energy, as shown in Figure 2. The surface contours for the >1, >10, and >100 MeV trapped protons are depicted as red, blue, and green curves, respectively, in the panel. These contours highlight how the low-energy trapped proton component is distributed in the southern regions of the SAA-even superimposing to the outer belt, while higher energy populations are more clearly enclosed in the classical boundaries of the inner belt, i.e., in the area above Brazil and the Atlantic Ocean.

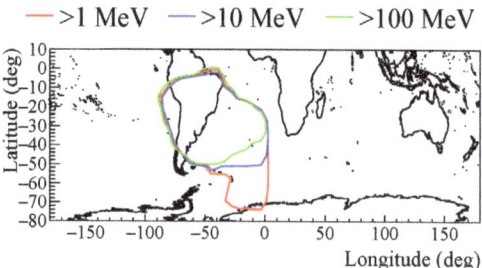

Figure 2. Geographical extension of the SAA for >1, >10, and >100 MeV protons (respectively, blue, red, and green curves in the panel), obtained from the AP9 model.

For a further, more precise comparison with experimental HEPD data, four different 20 min orbit portions were selected among those crossing the SAA, in order to build the related time-profiles of protons at various energies (With a period of 94.6 min, the satellite makes ~15 complete orbits per day.):

- orbit 1-13:26:00/13:54:00
- orbit 2-15:02:00/15:20:00
- orbit 3-16:32:00/16:56:00
- orbit 4-18:07:00/18:29:00

These passages over the SAA are represented in Figure 3 as a function of the geographical coordinates; while orbit 1 crosses the SAA in the outermost and peripheral region, orbit 3 crosses the bulk of the Anomaly, where particle fluxes are expected to be higher.

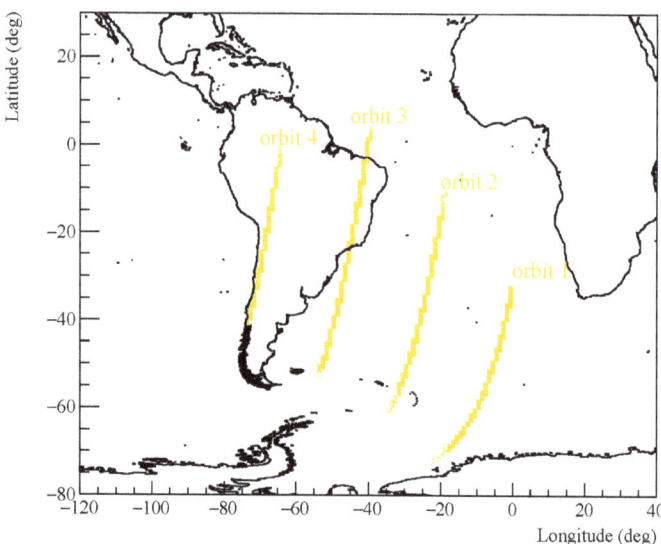

Figure 3. Representation of the four orbits chosen for the time-profile evaluation as a function of geographical latitude and longitude; orbit 1 appears to be more peripheral with respect to, for example, orbit 3.

5. Results

The differential, omnidirectional energy spectra of trapped protons along the four portions of orbits depicted in Figure 3, all generated by the AP9 model, are shown in Figure 4. As energy increases, the spectrum in each orbit decreases with a somewhat different steepness, as expected. This is, in fact, due to the different aspects of the trapping mechanism, which is the resulting effects of cosmic ray albedo neutron decay (CRAND), solar proton injection, and radial diffusion [45]. The CRAND mechanism is the principal trapped proton source above ~100 MeV, and the shape of the albedo neutron vertical spectrum above the geomagnetic cut-off is very similar to the one observed in trapped protons, i.e., the spectrum is decreasing as energy increases. On the contrary, the solar injection is more relevant below ~100 MeV (and it is more important for L > 2), while the radial diffusion tends to redistribute trapped particles in different L, so its effects are more complex. As a result, the trapped proton flux is strongly anisotropic and the overall spectrum changes rapidly, heavily depending on the region (latitude, longitude, L, etc.) where it is estimated.

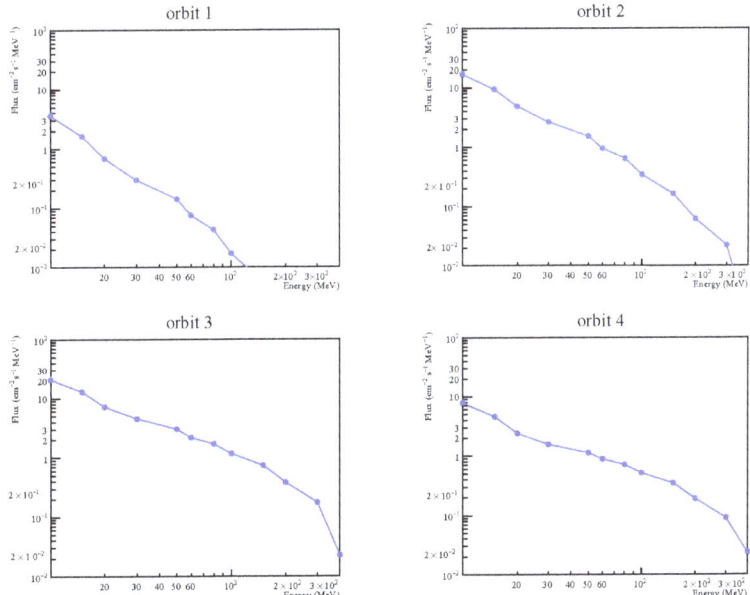

Figure 4. Differential, omnidirectional energy spectra of >10 MeV trapped protons obtained with the AP9 model and averaged over each CSES-01 orbit (see title above each panel). In each orbit, the spectrum decreases as energy increases, as expected.

The time profiles (5-s resolution) for the >10 MeV trapped protons along the orbits defined above are also shown in Figure 5. In each panel, the color palette relates to the different particle energy. Note that during each portion of the CSES-01 orbits, the spectra possess a different shape as a function of time. For example, during orbit 1, trapped fluxes tend to decrease very rapidly, while for the other orbits this decrease is slower. This is due to the fact that orbit 1 crosses the SAA region only in its peripheral section, so the trapped population is only encountered for a small amount of time. Furthermore, in each panel it seems that energetic protons are more concentrated in the internal sectors of the SAA, while low-energy protons are more widely distributed and spread over a larger area; this was inferred also from Figure 2, and it is another proof of the high variability of trapped fluxes inside the inner belt.

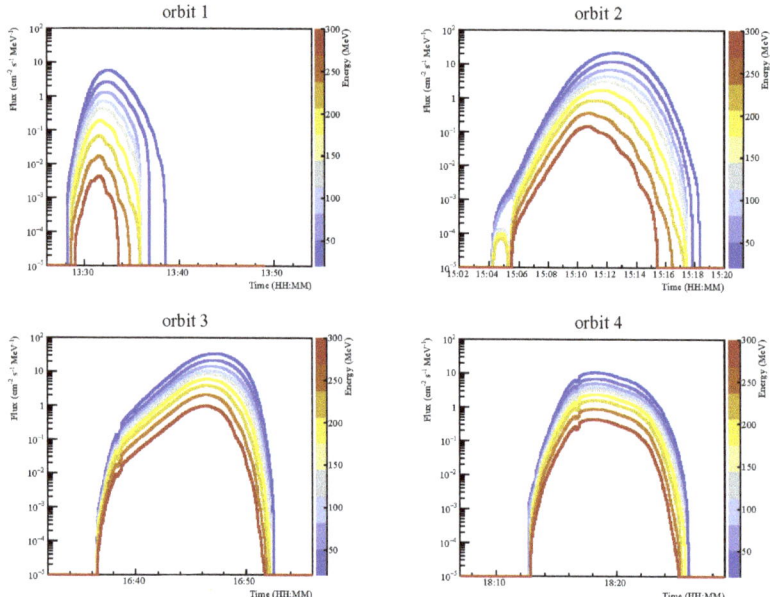

Figure 5. Time profiles (5-s resolution) of 10–300 MeV trapped protons estimated from the AP9 model along CSES orbits 1–4 (from the top left panel to the lower right panel). As expected, higher energies have lower fluxes, while lower energies tend to have higher fluxes.

To assess the level of agreement between the AP9 model and the experimental data, a preliminary analysis was conducted using omnidirectional ∼50 MeV calibrated proton data from HEPD, obtained following the same procedure used in [40]. As can be seen from the four panels in Figure 6, the agreement seems to be good, even if some small discrepancies are evident, mostly in the peripheral regions of the SAA; these are probably due to the different operational definition of South Atlantic Anomaly that was used to derive the data with HEPD (For the HEPD data analysis, we define the South Atlantic Anomaly as the region enclosed in a value of the magnetic field >20,000 nT.). Further studies are needed to verify the agreement even in a longer time period and with the extensive use of simulations.

For the purposes of this preliminary analysis, the uncertainties of the AP9 model—related to measurement, gap-filling, or dynamic variations due to space-weather processes—are not taken into account. A future, more complete comparison with HEPD observations will require a precise assessment of the AP9 confidence levels, in order to better evaluate the match with experimental data. Considering that CSES-01 will be operative in a period of strong minimum between the end of the 24th solar cycle and the start of the 25th, no major effect related to space weather variability is expected.

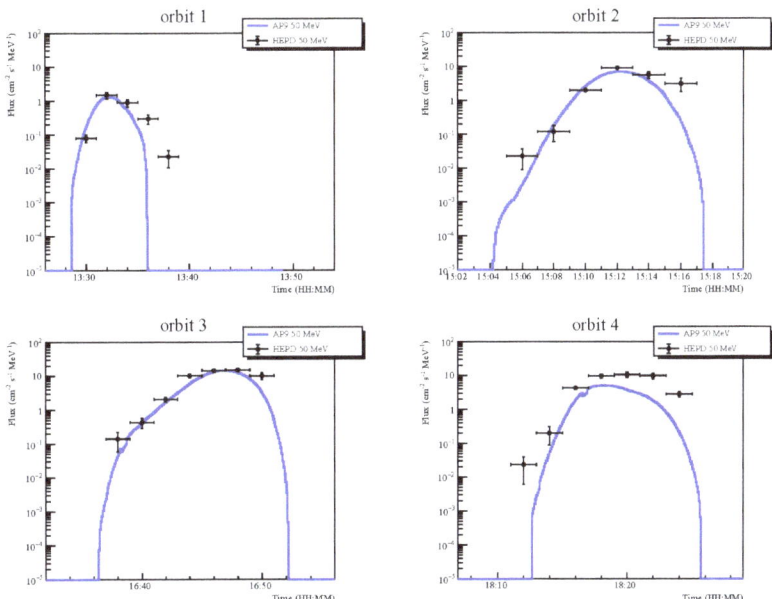

Figure 6. Time profiles (5-s resolution) of 50 MeV trapped protons estimated from the Ap9 model and compared with preliminary data of ~50 MeV proton data (black circles) from the HEPD instrument on board the CSES-01 satellite. The analysis has been carried out using the procedure described in [40]. The agreement between the data and the model appears generally good, despite showing small discrepancies, especially in the peripheral regions of the SAA. Only statistical errors are reported.

6. Discussion and Conclusions

The NASA AE9/AP9/SPM set of models represents an important approach to specify the radiation environment for modern satellite design applications. In this work, this suite of models has been employed to estimate trapped proton fluxes over the South Atlantic Anomaly for some orbits of the China Seismo-Electromagnetic Satellite on 1 January 2021. This is intended as a starting point for a future analysis that will include data from the High-Energy Particle Detector. After three years of calibration and testing, HEPD has proven capable of measuring low-energy particle fluxes (>3 MeV electrons and >35 MeV protons) with precision and stability over time; among the others, these two characteristics in particular are very suitable for the measurement of strongly anisotropic particle fluxes, such as those trapped in SAA. Thus, HEPD, together with the other payloads on board CSES (such as those of the HEPP suite), can provide excellent cross-calibration for these radiation environment models at LEO. A preliminary analysis on HEPD proton data has been conducted to assess the agreement between the AP9 model and experimental data, and it seems already acceptable, even if some discrepancies—that need to be studied—are present. It is important to remember, as already mentioned, that there is also a certain number of known issues in these models:

- There are no reliable data for inner region electrons at energies <1 MeV and spectral/spatial extrapolation of the few existing datasets can lead to large deviations.
- There are no data for high-energy protons (>150 MeV). AP9 goes out to 400 MeV only by using physics-based model extrapolation techniques.

Moreover, much of the validation of these models was performed using the Van Allen Probe mission [46], which provided a rich set of energetic particle and plasma data from the many instruments the spacecraft carried on board, together with a good pitch angle and energy resolution; unfortunately, after the end of the mission, new data are necessary to continue validation and to explore higher energy ranges with more statistics. HEPD proved to be able to cover this role, performing measurements with precision and stability in time; besides, new CSES missions (with more HEPD-like particle detectors) are already planned for the next years, greatly expanding the data-taking period by several years into the 2020s.

Author Contributions: Writing original draft, M.M. (Matteo Martucci); Conceptualization, R.S; Writing-review and editing, R.S., R.B., L.C. (Livio Conti), F.M.F., M.P. (Mirko Piersanti), A.P., C.D.D., A.S., F.P. (Francesco Palma); Designing the experiment or calibration or data production and processing, S.B., R.B., W.J.B., D.C., L.C. (Luca Carfora), G.C., L.C. (Livio Conti), A.C., C.D.D., C.D.S., F.M.F., R.I., I.L., N.M., G.M., M.M. (Matteo Mergé), A.O., G.O., F.P. (Francesco Palma), F.P. (Federico Palmonari), B.P., A.P., F.P. (Francesco Perfetto), P.P., M.P. (Mirko Piersanti), M.P. (Michele Pozzato), E.R., M.R., S.B.R., Z.S., V.S., A.S., V.V., S.Z., P.Z. All authors have read and agreed to the published version of the manuscript.

Funding: This research received no external funding.

Institutional Review Board Statement: Not applicable.

Informed Consent Statement: Not applicable.

Data Availability Statement: CSES/HEPD data can be found in www.leos.ac.cn/, accessed on 1 February 2021.

Acknowledgments: This work makes use of data from the CSES mission (www.leos.ac.cn/, accessed on 1 February 2021), a project funded by China National Space Administration (CNSA), China Earthquake Administration (CEA) in collaboration with the Italian Space Agency (ASI), National Institute for Nuclear Physics (INFN), Institute for Applied Physics (IFAC-CNR), and Institute for Space Astrophysics and Planetology (INAF-IAPS). We kindly acknowledge the AFRL for providing the AE9/Ap9/SPM set of models. This work was supported by the Italian Space Agency in the framework of the "Accordo Attuativo 2020-32.HH.0 Limadou Scienza+" (CUP F19C20000110005) and the ASI-INFN agreement n.2014-037-R.0, addendum 2014-037-R-1-2017.

Conflicts of Interest: The authors declare no conflict of interest.

Abbreviations

The following abbreviations are used in this manuscript:

CSES	China Seismo-Electromagnetic Satellite
HEPD	High-Energy Particle Detector
SAA	South Atlantic Anomaly
AFRL	Air Force Research Laboratory
TLE	Two-Line Elements
SGP4	Simplified General Perturbations
CRAND	Cosmic Ray Albedo Neutron Decay
SEE	Single Event Effect
NRO	National Reconnaissance Office
VDL	Virtual Distributed Laboratory
LEO	Low-Earth Orbit
ELSET	Element Set

References

1. Cummings, J.R.; Cummings, A.C.; Mewaldt, R.A.; Selesnick, R.S.; Stone, E.C.; von Rosenvinge, T.T. New Evidence for Anomalous Cosmic Rays Trapped in the Magnetosphere. In Proceedings of the 23rd International Cosmic Ray Conference (ICRC23), Calgary, AB, Canada, 19–30 July 1993; Volume 3, p. 428.
2. Adriani, O.; Barbarino, G.C.; Bazilevskaya, G.A.; Bellotti, R.; Boezio, M.; Bogomolov, E.A.; Bongi, M.; Bonvicini, V.; Bottai, S.; Bruno, A.; et al. Trapped proton fluxes at low Earth orbits measured by the PAMELA experiment. *Astrophys. J.* **2015**, *799*, L4. [CrossRef]
3. Singer, S.F. Trapped Albedo Theory of the Radiation Belt. *Phys. Rev. Lett.* **1958**, *1*, 181–183. [CrossRef]
4. Farley, T.A.; Walt, M. Source and loss processes of protons of the inner radiation belt. *J. Geophys. Res.* **1971**, *76*, 8223. [CrossRef]

5. Van Allen, J.A.; McIlwain, C.E.; Ludwig, G.H. Radiation Observations with Satellite 1958 ε. *JGR* **1959**, *64*, 271–286. [CrossRef]
6. Van Allen, J.A.; Frank, L.A. Radiation Measurements to 658,300 Km. with Pioneer IV. *Nature* **1959**, *184*, 219–224. [CrossRef]
7. Wrenn, G.L.; Sims, A.J. Internal Charging in the Outer Zone and Operational Anomalies. *Wash. DC Am. Geophys. Union Geophys. Monogr. Ser.* **1996**, *97*, 275. [CrossRef]
8. Brautigam, D.H. CRRES in review: Space weather and its effects on technology. *J. Atmos. Sol.-Terr. Phys.* **2002**, *64*, 1709–1721. [CrossRef]
9. Ginet, G.P.; O'Brien, T.P.; Huston, S.L.; Johnston, W.R.; Guild, T.B.; Friedel, R.; Lindstrom, C.D.; Roth, C.J.; Whelan, P.; Quinn, R.A.; et al. AE9, AP9 and SPM: New Models for Specifying the Trapped Energetic Particle and Space Plasma Environment. *Space Sci. Rev.* **2013**, *179*, 579–615. [CrossRef]
10. Ripa, J.; Dilillo, G.; Campana, R.; Galgoczi, G. A comparison of trapped particle models in low Earth orbit. In Proceedings of the Space Telescopes and Instrumentation 2020: Ultraviolet to Gamma Ray, Online, 13 December 2020. [CrossRef]
11. Fung, S.F.; Boscher, D.M.; Bilitza, D.; Tan, L.C.; Cooper, J.F. Modelling the Low-Altitude Trapped Radiation Environment. In Proceedings of the Environment Modeling for Space-Based Applications, Noordwijk, The Netherlands, 18–20 September 1996; Volume 392, p. 65.
12. Daly, E.J.; Evans, H.D.R. Problems in radiation environment models at low altitudes. *Radiat. Meas.* **1996**, *26*, 363–368. [CrossRef]
13. Brautigam, D.H.; Ray, K.P.; Ginet, G.P.; Madden, D. Specification of the Radiation Belt Slot Region: Comparison of the NASA AE8 Model With TSX5/CEASE Data. *IEEE Trans. Nucl. Sci.* **2004**, *51*, 3375–3380. [CrossRef]
14. Heirtzler, J. The future of the South Atlantic Anomaly and implications for radiation damage in space. *J. Atmos. Sol.-Terr. Phys.* **2002**, *64*, 1701–1708. [CrossRef]
15. Gubbins, D.; Jones, A.L.; Finlay, C.C. Fall in Earth's magnetic field is erratic. *Science* **2006**, *312*, 900–902. [CrossRef]
16. Aubert, J. Geomagnetic forecasts driven by thermal wind dynamics in the Earth's core. *Geophys. Suppl. Mon. Not. R. Astron. Soc.* **2015**, *203*, 1738–1751. [CrossRef]
17. Aubert, J.; Aurnou, J.; Wicht, J. The magnetic structure of convection-driven numerical dynamos. *Geophys. J. Int.* **2008**, *172*, 945–956. [CrossRef]
18. Wicht, J.; Christensen, U.R. Torsional oscillations in dynamo simulations. *Geophys. J. Int.* **2010**, *181*, 1367–1380. [CrossRef]
19. Gauß, C.F. Anzeige der Abhandlung des Herrn Hofr. Gauß: Intensitas vis magneticae terrestris admensuram absolutam revocata. *Astron. Nachrichten* **1833**, *10*, 349.
20. Erwan, T.; Finlay, C.; Beggan, C.; Alken, P.; Aubert, J.; Barrois, O.; Bertrand, F.; Bondar, T.; Boness, A.; Brocco, L.; et al. International Geomagnetic Reference Field: The 12th generation. *Earth Planets Space* **2015**, *67*, 79. [CrossRef]
21. Olsen, N.; Haagmans, R. Swarm-The earth's magnetic field and environment explorers. *Earth Planets Space* **2006**, *58*, 349–495. [CrossRef]
22. Pavón-Carrasco, F.J.; De Santis, A. The South Atlantic Anomaly: The Key for a Possible Geomagnetic Reversal. *Front. Earth Sci.* **2016**, *4*, 40. [CrossRef]
23. Deme, S.; Reitz, G.; Apáthy, I.; Héjja, I.; Láng, E.; Fehér, I. Doses Due to the South Atlantic Anomoly During the Euromir'95 Mission Measured by an On-Board TLD System. *Radiat. Prot. Dosim.* **1999**, *85*, 301–304. [CrossRef] [PubMed]
24. De Santis, A.; Qamili, E.; Wu, L. Toward a possible next geomagnetic transition? *Nat. Hazards Earth Syst. Sci.* **2013**, *13*, 3395–3403. [CrossRef]
25. Shen, X.; Zhang, X.; Yuan, S.; Wang, L.; Cao, J.; Huang, J.; Zhu, X.; Piergiorgio, P.; Dai, J. The state-of-the-art of the China Seismo-Electromagnetic Satellite mission. *Sci. China E Technol. Sci.* **2018**, *61*, 634–642. [CrossRef]
26. Piersanti, M.; Materassi, M.; Battiston, R.; Carbone, V.; Cicone, A.; D'Angelo, G.; Diego, P.; Ubertini, P. Magnetospheric-Ionospheric-Lithospheric Coupling Model. 1: Observations during the 5 August 2018 Bayan Earthquake. *Remote Sens.* **2020**, *12*, 3299. [CrossRef]
27. Piersanti, M.; De Michelis, P.; Del Moro, D.; Tozzi, R.; Pezzopane, M.; Consolini, G.; Marcucci, M.F.; Laurenza, M.; Di Matteo, S.; Pignalberi, A.; et al. From the Sun to Earth: Effects of the 25 August 2018 geomagnetic storm. *Ann. Geophys.* **2020**, *38*, 703–724. [CrossRef]
28. Varotsos, P.A.; Sarlis, N.V.; Skordas, E.S. Self-organized criticality and earthquake predictability: A long-standing question in the light of natural time analysis. *EPL (Europhys. Lett.)* **2020**, *132*, 29001. [CrossRef]
29. Varotsos, P.A.; Sarlis, N.V.; Skordas, E.S. Study of the temporal correlations in the magnitude time series before major earthquakes in Japan. *J. Geophys. Res. Space Phys.* **2014**, *119*, 9192–9206. [CrossRef]
30. Sarlis, N.V.; Skordas, E.S.; Varotsos, P.A.; Nagao, T.; Kamogawa, M.; Tanaka, H.; Uyeda, S. Minimum of the order parameter fluctuations of seismicity before major earthquakes in Japan. *Proc. Natl. Acad. Sci. USA* **2013**, *110*, 13734–13738. [CrossRef] [PubMed]
31. Li, X.Q.; Xu, Y.B.; An, Z.H.; Liang, X.H.; Wang, P.; Zhao, X.Y.; Wang, H.Y.; Lu, H.; Ma, Y.Q.; Shen, X.H.; et al. The high-energy particle package onboard CSES. *Radiat. Detect. Technol. Methods* **2019**, *3*, 22. [CrossRef]
32. Picozza, P.; Battiston, R.; Ambrosi, G.; Bartocci, S.; Basara, L.; Burger, W.J.; Campana, D.; Carfora, L.; Casolino, M.; Castellini, G.; et al. Scientific Goals and In-orbit Performance of the High-energy Particle Detector on Board the CSES. *Astrophys. J. Suppl.* **2019**, *243*, 16. [CrossRef]
33. Cheng, B.; Zhou, B.; Magnes, W.; Lammegger, R.; Pollinger, A. High precision magnetometer for geomagnetic exploration onboard of the China Seismo-Electromagnetic Satellite. *Sci. China Technol. Sci.* **2018**, *61*, 659. [CrossRef]

34. Cao, J.; Zeng, L.; Zhan, F.; Wang, Z.; Wang, Y.; Chen, Y. Meng, Q.; Ji, Z.; Wang, P.; Liu, Z.; et al. The electromagnetic wave experiment for CSES mission: Search coil magnetometer. *Sci. China Technol. Sci.* **2018**, *61*, 653. [CrossRef]
35. Diego, P.; Bertello, I.; Candidi, M.; Mura, A.; Coco, I.; Vannaroni, G.; Ubertini, P.; Badoni, D. Electric field computation analysis for the Electric Field Detector (EFD) on board the China Seismic-Electromagnetic Satellite (CSES). *Adv. Space Res.* **2017**, *60*, 2206–2216. [CrossRef]
36. Lin, J.; Shen, X.; Hu, L.; Wang, L.; Zhu, F. CSES GNSS ionospheric inversion technique, validation and error analysis. *Sci. China Technol. Sci.* **2018**, *61*, 669. [CrossRef]
37. Yan, R.; Guan, Y.; Shen, X.; Huang, J.; Zhang, X.; Liu, C.; Liu, D. The Langmuir Probe onboard CSES: Data inversion analysis method and first results. *Earth Planet. Phys.* **2018**, *2*, 479–488. [CrossRef]
38. Chen, L.; Ou, M.; Yuan, Y.; Sun, F.; Yu, X.; Zhen, W. Preliminary observation results of the Coherent Beacon System onboard the China Seismo-Electromagnetic Satellite-1. *Earth Planet. Phys.* **2018**, *2*, 505–514. [CrossRef]
39. Liu, C.; Guan, Y.; Zheng, X.; Zhang, A.; Piero, D.; Sun, Y. The technology of space plasma in-situ measurement on the China Seismo-Electromagnetic Satellite. *Sci. China Technol. Sci.* **2019**, *62*, 829–838. [CrossRef]
40. Bartocci, S.; Battiston, R.; Burger, W.J.; Campana, D.; Carfora, L.; Castellini, G.; Conti, L.; Contin, A.; Donato, C.D.; Persio, F.D.; et al. Galactic Cosmic-Ray Hydrogen Spectra in the 40–250 MeV Range Measured by the High-energy Particle Detector (HEPD) on board the CSES-01 Satellite between 2018 and 2020. *Astrophys. J.* **2020**, *901*, 8. [CrossRef]
41. Martucci, M.; Ambrosi, G.; Battiston, R.; Bartocci, S.; Basara, L.; Burger, W.; Campana, D.; Carfora, L.; Casolino, M.; Castellini, G.; et al. Space-Weather capabilities and preliminary results of the High Energy Particle Detector (HEPD) on-board the CSES-01 satellite. In Proceedings of the 36th International Cosmic Ray Conference—PoS(ICRC2019), Madison, WI, USA, 24 July–1 August 2019; Volume 358, p. 1118. [CrossRef]
42. Ambrosi, G.; Bartocci, S.; Basara, L.; Battiston, R.; Burger, W.J.; Campana, D.; Carfora, L.; Castellini, G.; Cipollone, P.; Conti, L.; et al. Beam test calibrations of the HEPD detector on board the China Seismo-Electromagnetic Satellite. *Nucl. Instrum. Methods Phys. Res. A* **2020**, *974*, 164170. [CrossRef]
43. Sotgiu, A.; De Donato, C.; Fornaro, C.; Tassa, S.; Scannavini, M.; Iannaccio, D.; Ambrosi, G.; Bartocci, S.; Basara, L.; Battiston, R.; et al. Control and data acquisition software of the high-energy particle detector on board the China Seismo-Electromagnetic Satellite space mission. *Softw. Pract. Exp.* **2020**, 1–22. [CrossRef]
44. McIlwain, C.E. Magnetic Coordinates. *Space Sci. Rev.* **1966**, *5*, 585–598. [CrossRef]
45. Selesnick, R.S.; Looper, M.D.; Mewaldt, R.A. A theoretical model of the inner proton radiation belt. *Space Weather.* **2007**, *5*. [CrossRef]
46. Mazur, J.; Friesen, L.; Lin, A.; Mabry, D.; Katz, N.; Dotan, Y.; George, J.; Blake, J.B.; Looper, M.; Redding, M.; et al. The Relativistic Proton Spectrometer (RPS) for the Radiation Belt Storm Probes Mission. *Space Sci. Rev.* **2013**, *179*, 221–261. [CrossRef]

Article
Low-Pass Filtering Method for Poisson Data Time Series

Victor Getmanov [1,2,*], Vladislav Chinkin [1], Roman Sidorov [1,*], Alexei Gvishiani [1,2], Mikhail Dobrovolsky [1], Anatoly Soloviev [1,2], Anna Dmitrieva [1,3], Anna Kovylyaeva [1,3], Nataliya Osetrova [1,3] and Igor Yashin [1,3]

1. Geophysical Center of the Russian Academy of Sciences (GC RAS), 119296 Moscow, Russia; v.chinkin@gcras.ru (V.C.); adg@gcras.ru (A.G.); m.dobrovolsky@gcras.ru (M.D.); a.soloviev@gcras.ru (A.S.); andmitriyeva@mephi.ru (A.D.); aakovylyaeva@mephi.ru (A.K.); nvosetrova@mephi.ru (N.O.); iiyashin@mephi.ru (I.Y.)
2. Schmidt Institute of Physics of the Earth of the Russian Academy of Sciences (IPE RAS), 123242 Moscow, Russia
3. Scientific & Educational Centre NEVOD, National Research Nuclear University MEPhI (NRNU MEPhI), 115409 Moscow, Russia
* Correspondence: v.getmanov@gcras.ru (V.G.); r.sidorov@gcras.ru (R.S.)

Abstract: Problems of digital processing of Poisson-distributed data time series from various counters of radiation particles, photons, slow neutrons etc. are relevant for experimental physics and measuring technology. A low-pass filtering method for normalized Poisson-distributed data time series is proposed. A digital quasi-Gaussian filter is designed, with a finite impulse response and non-negative weights. The quasi-Gaussian filter synthesis is implemented using the technology of stochastic global minimization and modification of the annealing simulation algorithm. The results of testing the filtering method and the quasi-Gaussian filter on model and experimental normalized Poisson data from the URAGAN muon hodoscope, that have confirmed their effectiveness, are presented.

Keywords: Poisson data; time series; quasi-Gaussian filter; digital filtering; optimization; global minimization; annealing simulation algorithm

1. Introduction

The article proposes a low-pass filtering method for Poisson-distributed data time series, based on a specially developed digital low-pass filter with finite impulse response (FIR filter), with gain equal to one at zero frequencies and non-negative weighting factors.

Here, low-pass filtering is applied in order to reduce noise in Poisson-distributed data to ensure the recognition of emerging fluctuations of mathematical expectations in them. Poisson-distributed, or Poisson data are found in various physical systems, for example, related to the heliosphere and magnetosphere of the Earth; the fluctuations of mathematical expectations of these data may contain information regarding the structures and characteristics of these systems.

A particular feature of the Poisson data origin is that they contain sufficient noises; it is known, for example, from [1] that their variance is numerically equal to mathematical expectation. Noise reduction in Poisson data can be achieved using common FIR filters [2,3], to which, within the framework of this article, we refer the filters based on commonly used windowing techniques, frequency sampling and inverse Fourier transforms [4,5]. However, there are a number of scientific and technical problems for which their application is not fully effective, for example, (1) recognition of small (in size and duration) mathematical expectation fluctuations in Poisson datasets; (2) digital processing of Poisson data with small mathematical expectation values.

Common FIR filters can potentially be used for the mentioned tasks, and their synthesis can be implemented according to given dimensions and cutoff frequencies. The synthesis procedures for common FIR filters are, in essence, the variants of approximation procedures for the specified species frequency response (FR) types; the accuracy of the FR approximations depends on the specified dimensions for the synthesized filters. Obviously, at large dimensions, the accuracy of these approximations is high and the errors in the resulting cutoff frequencies are small. For the case of small dimensions, the approximation accuracy turns out to be low and, as a consequence, cutoff frequencies are realized with significant errors which prevent low-pass filtering. We can assume that the filtering procedure proposed here should be performed by filters with low dimensions and cutoff frequencies and with gain values equal to one in order to avoid mathematical expectation distortions, and with non-negative weight factors in order to provide non-negativity of filtering results taking into account the Poisson property of the data.

The indicated problem leads to the need to formulate the synthesis problem for a special digital low-pass FIR filter, which takes into account the requirements—restrictions on dimensionality, cutoff frequency, gain at zero frequencies, and weighting factors.

Here, a FIR filter is proposed, which is further denoted as a quasi-Gaussian filter, the frequency response of which is formed on the basis of approximating a Gaussian function and ensuring the implementation of the mentioned constraints conditions using a special optimization method.

Gaussian filters, the frequency response of which is implemented based on the approximation of the Gaussian function, are widely used in modern scientific and technical practice [6,7]. However, as a rule, the known variants of Gaussian filters with the approximation of the frequency response do not take into account the above-mentioned conditions (restrictions).

Problems of digital processing of Poisson data time series from muon counters in muon detectors and telescopes [8], counters of elementary particles of alpha-beta-gamma radiation, photon counters, slow neutrons, etc. [9], taking into account their specificity, are relevant for experimental physics. Digital processing of Poisson data, including the Gaussian filtering application, can be outside of experimental physics, for example, in medical technology for imaging blood vessels and tumor therapy with particle beams, in measuring technology for tribological studies of the surfaces of metal parts, in astronomy for gamma telescopes, in muon tomography for recognizing cavities in rocks, and building structures and many other applications.

One of the applications of the designed filter proposed here is the digital processing of the data from the URAGAN muon hodoscope (MH) designed by NRNU MEPhI [10,11]. The MH is a computerized measuring device that estimates the intensities of muon fluxes by counting the number of elementary particles—muons—registered in its detector for a set of solid angles with a set time step. Within the framework of this article, MH can be interpreted as a distributed set of muon counters, consisting of primary and secondary information converters.

From each primary MH transducer, the initial Poisson data—time series of random non-negative integers $N(Tk)$—the quantities of Poisson-distributed events recorded in a given solid angle at time intervals $(T(k-1),\ T(k-1)+T_{0k})$, $k=1,2,\ldots,k_0$, where $T=1$ minute. Due to the features of the MH design, registration intervals T_{0k} are random with a uniform distribution law in the range $T_{0\min} \leq T_{0k} \leq T_{0\max} < T$.

From each secondary MH transducer, the 1-minute-sampled normalized Poisson data $Y(Tk)$ are generated for a given solid angle by reducing to one second and calculating the averaged normalized Poisson data $Y(T_0n)$ with an hourly discreteness according to the following relations:

$$Y(Tk) = N(Tk)/T_{0k}, Y(T_0n) = \frac{1}{60} \sum_{k=1+60(n-1)}^{k=60n} Y(Tk), n = 1,2,\ldots, T_0 = T \cdot 60. \quad (1)$$

Data resulting from (1) are produced for the whole set of solid angles; next, they are placed into time series of matrix MH data in the database [12].

2. Method

2.1. Quasi–Gaussian Digital Low-Pass Filter

2.1.1. Statement of the Problem

One-dimensional FIR filter synthesized here is built according to the following difference equation:

$$X(T_0 n) = \sum_{s=0}^{s_0} a_s Y(T_0(n-s)), n = 1, 2, \ldots,\qquad(2)$$

where $r_0 = s_0 + 1$ is the FIR filter dimension, $a^T = (a_0, a_1, \ldots, a_{s_0})$ is a weight factors vector, $X(T_0 n)$ is the output time series, $Y(T_0 n)$ is the FIR filter input—the hourly normalized Poisson data time series from MH according to (2), which begins from the values $Y(T_0(1-s_0))$, $Y(T_0(1-s_0+1))$, $Y(T_0(1-s_0+2))$,.... Transfer function (TF) $H(j\omega T_0, a)$ for filter (2) is defined as follows:

$$H(j\omega T_0, a) = \sum_{s=0}^{s_0} a_s e^{-j2\pi \omega T_0 s}.\qquad(3)$$

Here ω is the TF frequency parameter. For (3), a normalized fequence is introduced, $w, \omega T_0 = w\pi, 0 \leq w \leq 1.0$, and its discrete values are calculated: w_l

$$dw = 1.0/L_0,\ w_l = dw(l-1),\ l = 1, \ldots, L, L = L_0 + 1.\qquad(4)$$

The frequency response (FR) $H(w_l, a) = |H(jw_l, a)|$, considering (3), is the following:

$$H(w_l, a)^2 = H_1^2(w_l, a) + H_2^2(w_l, a),$$
$$H_1(w_l, a) = \sum_{s=0}^{s_0} a_s \cos(2\pi w_l s),\qquad(5)$$
$$H_2(w_l, a) = \sum_{s=0}^{s_0} a_s \sin(2\pi w_l s)$$

for discrete normalized frequencies w_l, $l = 1, \ldots, L$ according to (4). The cutoff frequency w_c for FR (5) is found based on the equality $|H(jw_A, a)|^2 = 0.5$.

For a low-frequency FIR filter synthesis, the FR of the prototype filter is used, based on a Gaussian function $H_{0g}(w, w_{c0})$

$$H_{0g}(w, w_{c0}) = \exp(-(w/w_{c0})^2).\qquad(6)$$

2.1.2. Synthesis Requiements

The problem of synthesis of the supposed FIR filter is solved based on the approximation of the FR function $H_{0g}(w_l, w_c)$ (6) in discrete points w_l, $l = 1, \ldots, L$ with a FR function $H_g(w_l, 0)$ according to (5). A functional $S(H_{0g}, a, w_c)$ is formed:

$$S(H_{0g}, a, w_c) = \sum_{l=1}^{L} [(\sum_{s=0}^{s_0} a_s C_s(w_l))^2 + (\sum_{s=0}^{s_0} a_s S_s(w_l))^2 - H_{0g}^2(w_l, w_c)]^2.\qquad(7)$$

Obviously, the FR (5) represents a function which is polyharmonic in frequency w_l. In case the prototype filter FR frequency derivative has discontinuities or is subject to strong alternations, e.g., if FR is a trapezoidal function, then the FR of the synthesized FIR filter, obtained based on approximation, will contain fluctuations due to the so-called Gibbs effect. Elimination and reduction of these fluctuations are usually achieved by choosing a suitable smooth prototype filter FR function. The smoothness requirement is largely satisfied by the Gaussian function (6). It should be noted that the Gaussian function is naturally suitable for the FR of a low-pass filter, since its values (6) practically differ from zero only in the region of low frequencies.

The requirements listed in the Introduction lead to formalized requirements:

a. Ensuring that the gain at zero frequencies is equal to one:

$$1 = H(0,a) = \sum_{s=0}^{s_0} a_s, a \in A_1, \ A_1 = \{a : (1 = \sum_{s=0}^{s_0} a_s)\}; \tag{8}$$

b. Ensuring non-negativity of coefficients:

$$a \in A_0, \ A_0 = \{a : (0 \leq a_s, s = 0,1,\ldots,s_0)\}; \tag{9}$$

For the synthesis procedure, it is assumed to set a small value r_0, based on the a priori known duration of fluctuations, and some small cutoff frequency value w_c for a prototype filter. The quasi-Gaussian filter synthesis procedure, consisting of finding the optimal coefficients $a_s^\circ, s = 0,1,\ldots,s_0$, taking into account the requirements **a**,**b**, Equations (8) and (9) the predefined r_0 and w_c, is performed on the basis of the approximation problem, which reduces to the implementation of conditional minimization:

$$a^\circ(w_c) = \arg\{\min_{a \in A_0, a \in A_1} S(H_{0g}, a, w_c)\}. \tag{10}$$

For a given small dimension r_0 of the synthesized quasi-Gaussian filter and a given small cutoff frequency w_c for a prototype filter, the value for cutoff frequency to be found for a quasi-Gaussian filter is w_{cg}, and the filter FR for the frequencies w_l is denoted as $H_g(w_l, w_{cg}, a^\circ), l = 1,\ldots,L$.

The minimization of (10) could be performed based on modified direct zero-order optimization methods, taking into account the restrictions (8) and (9). However, because the (7) functional is multi-extremal, traditional modified direct methods, for example, using the coordinate descent method, the Hook–Jeeves method, the random descent method, etc. [13] do not provide successful minimization. The listed methods, as a rule, lead to "getting stuck" with search procedures in local minima.

2.2. Quasi–Gaussian Filter Synthesis Procedure

We can synthesize the quasi-Gaussian filter based on the technology of stochastic global minimization of the (7) functional with the constraints (8) and (9) using the optimization algorithm for annealing simulation [14,15]. To implement it, we will use the simulannealbnd.mat software module from the Matlab Global Optimization Toolbox [16].

Let us form a parallelepiped of constraints $\overline{A}_0$ of dimension r_0 with boundaries $\overline{a}_r, r = 1,\ldots,r_0$—$a \in \overline{A}_0, \ \overline{A}_0 = \{a : (0 \leq a_r \leq \overline{a}_r, r = 1,\ldots,r_0)\}$ and a new—with respect to (7)—functional $\overline{S}(H_{0g}, a)$ with a penalty term taking into account the constraint equality (8). Let us implement the global minimization of $\overline{S}(H_{0g}, a)$ taking into account $\overline{A}_0$ using [16].

Let us set the initial vectors for the first iteration $a_1(I) \in \overline{A}_0$, uniformly distributed in $\overline{A}_0$, I—a single descent procedure, $I = 1,2,\ldots,I_0$, I_0—a total number of descent procedures. Let us assume that each descent procedure consists of m_0—a total number of iterations, m—a single iteration, $m = 1,2,\ldots,m_0$. During descent, we assume that the initial value of the vector of parameters for $(m+1)$-st iteration is equal to the calculated optimal value for the vector of parameters for m-th iteration—$a_{m+1}(I) = a_m^\circ(I)$. In each iteration, we perform n_0 descent steps, n is a descent step number, $n = 1,2,\ldots,n_0$. Next, we will calculate the sequence of the functional $S(m_0, I) = \overline{S}(H_{0g}, a^\circ)$ values and the corresponding optimal vectors $a^\circ(m_0, I), I = 1,2,\ldots,I_0$. For global minimization, we search for the optimal index I° corresponding to the minimum of the $S(m_0, I)$ functional, and the optimal vector a° using brute force:

$$I^\circ = \arg\{\min_{1 \leq I \leq I_0} S(m_0, I)\}, a^\circ = a^\circ(m_0, I^\circ).$$

On Figure 1, the example plots of the minimized $S(m, I)$ functionals are displayed, depending on iteration number m and the descent procedure number I. Functionals are shown starting with $m = 2$, since for $m = 1$ their values are very large. Here, $m_0 = 20$; as the iteration number increases, the values of the functionals decrease. During the optimization process, a movement is made in a r_0-dimensional space from one local minimum to another.

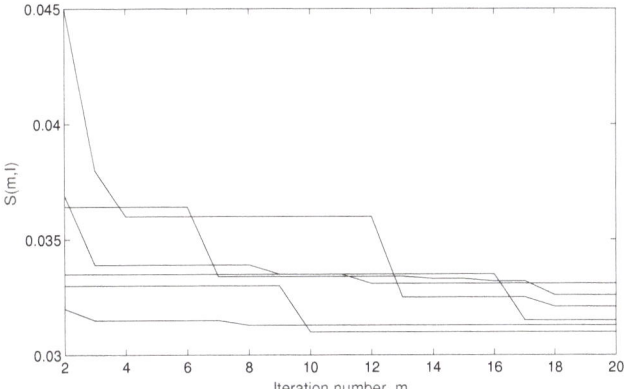

Figure 1. Plots of descent procedures—minimization of functionals $S(m, I)$, $I = 1, 2, \ldots, l_0$, $m = 1, 2, \ldots, m_0$.

Let us consider an example of quasi-Gaussian FIR filter synthesis. Based on the analysis of hourly experimental MH data from [12], it was found that the durations of possible fluctuations of the mathematical expectation in them were, on average, $\approx 10 \div 20$ h and more. The dimension value r_0, that could possibly allow the recognition of such fluctuations in mathematical expectations, was equal to 8. For a prototype filter FR (6), the parameter w_{c0} was related to the assigned cutoff frequency w_c based on (6)

$$(0.5)^{1/2} = \exp(-(w_c/w_{c0})^2), w_{c0} = w_c/(0.5 \cdot \ln 2)^{1/2}.$$

We assign the cutoff frequency $w_c = 0.1$, find w_{c0} and define $H_{0g}(w, w_c)$—the prototype filter FR. By defining L we set the number of discrete normalized frequencies w_l of calculations of the functional (7) for $0 \leq w_l \leq 1.0$, let us assume that $L = 100$ in our calculations. The polyharmonic FR function $|H(jw)|$ (5) is formed from components performing $1, 2, \ldots, s_0$ fluctuations in this interval. For the accepted values L and r_0, one period of the polyharmonic component with the maximum frequency corresponding to the number s_0 in (5), accounted for ≈ 15 sampling points of normalized frequencies w_l, $l = 1, \ldots, L$, which fully provided a fairly accurate calculation of the functional (7) necessary for direct search.

Let us calculate the vector of factors $a°$, form the synthesized quasi-Gaussian filter FR $H_g(w, w_{cg}, a°)$ and define the cutoff frequency $w_{cg} = 0.175$.

For the comparison, let us synthesize a common FIR filter using the fir1.mat module [3]. For the dimension $r_0 = 8$ and the assigned cutoff frequency $w_c = 0.1$ we find out the final cutoff frequency $w_{cf} = 0.275$; let us denote the FR as $H_f(w, w_{cf})$. On Figure 2, the FR plots for $H_{0g}(w, w_c)$, $H_g(w, w_{cg}, a°)$, $H_f(w, w_{cf})$ are displayed. It is seen that, in case of low r_0, the quasi-Gaussian filter FR was characterized by a better approximation to the prototype filter FR than the one of the common FIR filter.

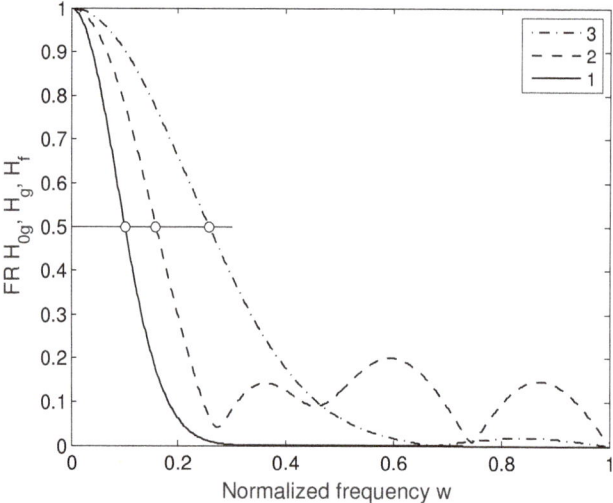

Figure 2. FR plots: $H_{0g}(w, w_c)$ (line 1), $H_g(w, w_{cg}, a°)$ (line 2), $H_f(w, w_{cf})$ (line 3).

Note that the proposed FIR filter, with the same dimension as the common FIR filter, made it possible to provide a lower value of the cutoff frequency than the realized cutoff frequency for the common FIR filter. The calculated cutoff frequencies of resulting FRs for common FIR filters synthesized using frequency sampling method and Fourier transforms [3,4] insignificantly (by ≈5–7%) differ from the cutoff frequency $w_{cf} = 0.275$. This gives a reason to make a conclusion about the advantages of a quasi-Gaussian filter over standard FIR filters.

3. Results

3.1. Testing the Method and the Quasi–Gaussian Filter on Model Normalized Poisson Data

3.1.1. Testing on Model Hourly Data Using Statistical Modeling

Testing of the proposed method and quasi-Gaussian filter was carried out on model hourly normalized data using statistical modeling [17]. For this purpose, on the basis of the Matlab module exprnd.mat [18], exponentially distributed model random numbers $\tau_i, i = 1, 2, \ldots$ were generated, with their mean value τ_{M0}, and the evenly distributed random registration time intervals $T_{0k}, k = 1, 2, \ldots$ within the range $T_{0\ min} \leq T_{0k} \leq T_{0\ max}$. The number of Poisson model events $N_M(Tk)$ was counted on the registration time intervals T_{0k}. Finding $N_M(Tk)$ was carried out by solving the conditional maximization problems:

$$N_M(Tk) = \arg\{\max N_M\}_i > 0, \quad (11)$$

providing that $T_{0k} - \sum_{i=1}^{N_M} \tau$, where for $T_{0k}, k = 1, 2, \ldots k_0$ the range bounds $T_{0\ min} = 57$ s, $T_{0\ max} = 59.5$ s were assigned (see the Introduction section). Initial model 1-minute-sampled and normalized Poisson-distributed data were constructed according to (11) and the calculation of relations $\overline{N}_M(Tk)$, similar to (1):

$$\overline{N}_M(Tk) = N_M(Tk)/T_{0k}, k = 1, 2, \ldots k_0. \quad (12)$$

The modulation of the average number of Poisson events in order to model decreases (increases) in the mathematical expectation was carried out by specifying the mean value function $\tau_{M0}(Tk)$ on the intervals $(T(k-1), Tk)$ for k from (12). For this, the relative modulation function $\mu(Tk), k = 1, 2, \ldots, k_0$ was formed and the initial temporal index of the modulation decrease k_a, the duration of the decrease dk_a and the depth of the relative decrease $d\mu$;. The function $\mu(Tk)$ was represented by the relations $\mu(Tk) = 1 - d\mu$ for $k_a \leq k \leq k_a + dk_a$, $\mu(Tk) = 1$ for $1 \leq k < k_a$, $k_a + dk_a + 1 \leq k \leq k_0$.

For the calculation example, the average number of Poisson model events per minute was set $N_{M0} = 25$, normalized average $\overline{N}_{M0} = N_{M0}/T$, modulated normalized mean $\overline{N}_{M0}(Tk) = \overline{N}_{M0}\mu(Tk) = N_{M0}\mu(Tk)/T$, $k = 1,2,\ldots,k_0$ and the parameter $\tau_{M0}(Tk) = 1/(\overline{N}_{M0}(Tk) - 1)$ was calculated.

Based on [18], random exponentially distributed numbers with $\tau_{M0}(Tk)$ and random evenly distributed values with $T_{0\ min} = 57s$, $T_{0\ max} = 59,5s$ were generated, with the use of which by (11), model Poisson data $N_M(Tk)$ and by (12)—normalized Poisson data $\overline{N}_M(Tk)$ were calculated. Further, similarly to (1), a time series of averaged model hourly normalized Poisson data was formed:

$$Y_M(T_0 n) = \frac{1}{60} \sum_{k=1+60(n-1)}^{k=60n} \overline{N}_M(Tk), n = 1,2,\ldots,n_0, n_0 = ent(k_0/60). \quad (13)$$

For modeling, we assumed $k_0 = 6000$, which corresponded to the model minute data produced during 4.166 days. For the modulation function, the values $k_a = 1920$, $dk_a = 1440$ and $d\mu = 0.02$ were taken. Model hourly averaged data $Y_M(T_0 n)$ for (13) with $n_0 = 100$, $n_{a1} = 32$, $n_{a2} = n_1 + dn_a\ dn_a = 24$.

Figure 3 shows an example of statistical modeling results: the jagged light gray line with index 1 displays the $Y_E(T_0 n)$ plot; the solid line with index 2 denotes the fragment of $X_{EG}(T_0 n)$ which is the result of filtering the model dataset using a quasi-Gaussian filter; for comparison, the dashed line with index 3 denotes the fragment $X_{EF}(T_0 n)$ which is the result of filtering the model dataset using the software module fir1.mat [3]. Model piecewise constant modulating function $Y_{M0}(T_0 n) = \overline{N}_M(T_0 n)$, represented by a dotted line (index 4), $m_0 + dm = Y_{M0}(T_0 n) = 0.4165$ for $1 \leq n < n_{a1}$, $n_{a2} \leq n < n_0$, $m_0 = Y_{M0}(T_0 n) = 0.4087$ for $n_{a1} \leq n \leq n_{a2}$, where the value of $dm = 0.833 \times 10^{-2}$ corresponded to the predefined 2% decrease. The plots show that the result of the quasi-Gaussian filter application (line 2) is a better approximation to the model piecewise constant modulation (line 4) than the result of a common FIR filtering (line 3).

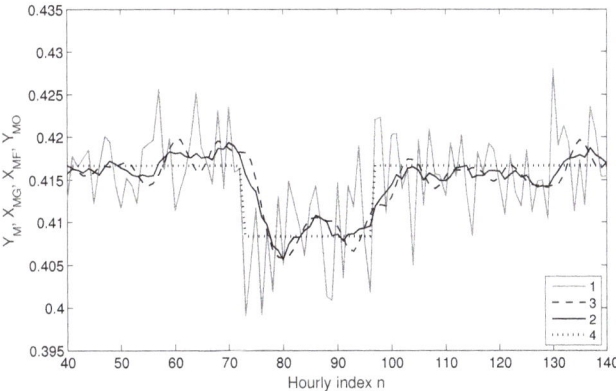

Figure 3. Fragments of model datasets $Y_M(T_0 n)$ (line 1), filtering results $X_{MG}(T_0 n)$ (line 2), $X_{MF}(T_0 n)$ (line 3) and model modulating function $Y_{M0}(T_0 n)$ (line 4).

The calculation of approximate estimates of filtering errors for the quasi-Gaussian filter and fir1-filter was performed by calculating the root-mean-square (RMS) errors according to the following formulas for datasets $Y_{M0}(T_0 n)$, $Y_{MG}(T_0 n)$, $Y_{MF}(T_0 n)$:

$$\sigma^2_{MG} = \frac{1}{n_0}\sum_{n=1}^{n_0}(Y_{M0}(T_0 n) - Y_{MG}(T_0 n))^2, \sigma^2_{MF} = \frac{1}{n_0}\sum_{n=1}^{n_0}(Y_{M0}(T_0 n) - Y_{MF}(T_0 n))^2. \quad (14)$$

Results of a large number of tests performed for (14) showed that the σ_{MG} error values for $X_{MG}(T_0i)$ regarding $Y_{MG}(T_0n)$ are, on average, 15–30% less than the corresponding σ_{MF} error values for $X_{MF}(T_0i)$. An overview of model $X_{MG}(T_0i)$ and $X_{MF}(T_0i)$ (Figure 3) made it possible to ensure that the minimum duration of the interval, within which recognition for the decrease $d\mu = 0.02$ can be performed, is 12–24 h.

The proposed method and the quasi-Gaussian filter provided more noise reduction than a common FIR filter. Consideration of the results of statistical modeling made it possible to draw a conclusion about the efficiency of the quasi-Gaussian filtering method.

3.1.2. Estimation of Mathematical Expectation and Its Root Mean Square Errors

Testing of the method and quasi-Gaussian filter for estimating the mathematical expectation and the RMS of its errors depending on dn_a—the duration of decreases and $d\mu$—the relative decrease value were carried out using statistical tests [17]. Random datasets $Y_M(s, T_0i)$, $X_{MG}(s, T_0i)$, $X_{MF}(s, T_0i)$, $s = 1, 2, \ldots, M$, where s is the number of the dataset, M is the total quantity of datasets. The estimates of mathematical expectation $m_g^\circ(dn_a, d\mu)$ and RMS values $\sigma_g^\circ(dn_a, d\mu)$ for $X_{MG}(s, T_0i)$ for a set of values dn_a and $d\mu$

$$m_g^\circ(s, dn_a, d\mu) = \frac{1}{n_a} \sum_{n=n_{a_1}}^{n_{a_1}+dn_a} X_{MG}(s, T_0n), m_g^\circ(dn_a, d\mu) = \frac{1}{M} \sum_{s=1}^{M} m_g^\circ(s, dn_a, d\mu),$$

$$\sigma_g^\circ(s, dn_a, d\mu) = \frac{1}{n_a - 1} \sum_{n=n_a}^{n_a+dn_a} (X_{MG}(s, T_0n) - m_g^\circ(s, dn_a, d\mu))^2, \quad (15)$$

$$\sigma_g^\circ(dn_a, d\mu) = \frac{1}{M} \sum_{s=1}^{M} \sigma_g^\circ(s, dn_a, d\mu).$$

The coefficients of relative errors $\varepsilon_{gm}^\circ(dn_a, d\mu)$, $\varepsilon_{g\sigma}^\circ(dn_a, d\mu)$ of the quasi-Gaussian filter as ratios of errors $m_g^\circ(dn_a, d\mu) - m_0$ and RMS $\sigma_g^\circ(s, dn_a, d\mu)$ to the values of dm reductions are the following:

$$\varepsilon_{gm}^\circ(dn_a, d\mu) = (m_g^\circ(dn_a, d\mu) - m_0)/dm, \varepsilon_{g\sigma}^\circ(dn_a, d\mu) = (\sigma_g^\circ(dn_a, d\mu))/dm \ldots \quad (16)$$

The coefficients ε_{gm}°, $\varepsilon_{g\sigma}^\circ$, calculated for $dn_a, d\mu$, characterized the recognition capabilities of quasi-Gaussian filtering model decreases. Similarly, using (15) and (16) $m_f^\circ(dn_a, d\mu)$ and $\sigma_f^\circ(dn_a, d\mu)$ for $X_{MF}(s, T_0n)$ and the coefficients $\varepsilon_{fm}^\circ(dn_a, d\mu)$, $\varepsilon_{f\sigma}^\circ(dn_a, d\mu)$. On Figure 4, the results of statistical tests are displayed, where $M = 500$. The $\varepsilon_{gm}^\circ(dn_a, d\mu)$ coefficients plots are the solid lines with indices 1, 2, and the $\varepsilon_{fm}^\circ(dn_a, d\mu)$ plots are the dashed lines with indices 3, 4. The coefficients ε_{gm}°, ε_{fm}° are given depending on the duration with the values $dn_a = 12, 24, 48, 72$ h and relative decreases in $d\mu$, taking the values of 0.01 (indices 1, 3) and 0.03 (indices 2, 4).

The effect of quasi-Gaussian filtering was determined based on the calculation of $\delta\varepsilon_{fg,m}^\circ$—the rates of errors with respect to the mathematical expectations:

$$\delta\varepsilon_{fg,m}^\circ(dn_a, d\mu) = (\varepsilon_{fm}^\circ(dn_a, d\mu) - \varepsilon_{gm}^\circ(dn_a, d\mu))/\varepsilon_{gm}^\circ(dn_a, d\mu) \quad (17)$$

The results of the $\delta\varepsilon_{fg,m}^\circ$ calculations according to (17) for some $d\mu$ and dn_a values are:

1. $\delta\varepsilon_{fg,m}^\circ = 0.115$ (11.5%) for $dn_a = 24$ and $d\mu = 0.01$;
2. $\delta\varepsilon_{fg,m}^\circ = 0.196$ (19,6%) for $dn_a = 24$ and $d\mu = 0.03$.

Analysis of the error values showed that the ε_{gm}° rate values appeared to be about 10–30% lower than the ε_{fm}° values. The nature of the dependencies of the estimates of the error coefficients for the $\varepsilon_{g\sigma}^\circ$ and $\varepsilon_{f\sigma}^\circ$ root mean square values for the same dn_a and $d\mu$ parameters is almost the same: the $\varepsilon_{g\sigma}^\circ$ are also ≈10–30% lower than the $\varepsilon_{f\sigma}^\circ$. This means that, for the recognition of decreases small in duration and magnitude, the use of a quasi-Gaussian filter is more preferable than the use of a common FIR filter.

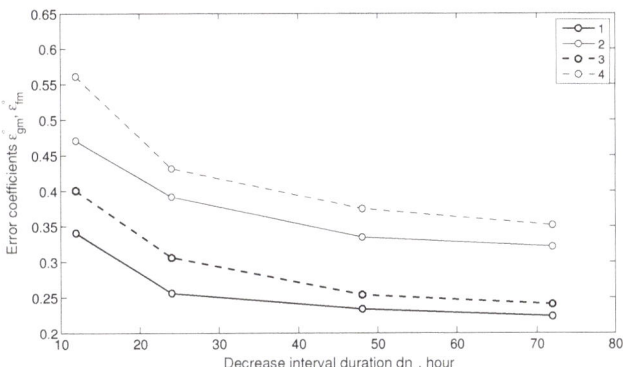

Figure 4. Results of calculating the coefficients of relative errors ε°_{gm}, ε°_{fm}.

3.2. Testing the Method and the Quasi–Gaussian Filter on Experimental Normalized Poisson Data from the URAGAN Hodoscope

Testing in this section consisted of determining the performance and capabilities of the proposed method and the quasi-Gaussian filter for recognizing small in duration and magnitude decreases in time intervals for the experimental hourly normalized Poisson data registered by the URAGAN hodoscope, taken from [12].

For analysis, a time interval was selected from 09/02/2017, 20:00 UTC to 09/18/2017, 15:00 UTC, with a total duration of 15.6 days. During this interval, the heliosphere was turbulent due to strong solar coronal mass ejections (CMEs) The CMEs that occurred on that period, caused intense geomagnetic storms that were discussed, for example, in [19,20]. The emerging CMEs caused modulations of muon fluxes recorded in MH and led to lower mathematical expectations (including the ones due to Forbush decreases) in Poisson MH data.

MH data were the matrix series of distribution functions of the intensities of muon fluxes $Y_E(i, j, T_0 n)$, defined in a rectangular region $i = 1, \ldots, N_1, j = 1, \ldots, N_2, N_1 = 90, N_2 = 76, n = 1, 2, \ldots$. Solid angles correspond to azimuth and zenith indices i, j, $\varphi_i = \Delta\varphi(i - 1)$, $\vartheta_j = \Delta\vartheta(j - 1)$, $\Delta\varphi = 1°$, $\Delta\vartheta = 4°$ in which the registered particles were counted. MH data $Y_E(j_0, i_0, T_0 n)$ were a time series with indices j_0, i_0; the considered interval was determined for $n_{E\min} \leq n \leq n_{E\max}$, $n_{E\min} = 5900$, $n_{E\max} = 6275$ (counting hours for [12] began from the first hour of 2017).

Figure 5 shows the results of quasi-Gaussian filtering and interval recognition with reductions in mathematical expectation. The original data $Y_E(T_0 n)$ for $j_0 = 30$, $i_0 = 31$ were denoted by light gray jagged lines (index 1). Fluctuations in data with a period of $\approx$24 h and an amplitude of $\approx$0.0037–0.0040 are due to the daily rotation of the MH with the Earth. Line with index 2 depicts the data $X_{EG}(T_0 n)$ filtered based on quasi-Gaussian filter. The recognized intervals of intensity decrease, intensity recovery and intensity mathematical expectation decrease were denoted by a piecewise linear spline-like dashed line $X_{ES}(T_0 n)$ (index 3). Analysis of intervals 5969–6043, 6127–6189 based on $X_{ES}(T_0 n)$ leads to a conclusion that the mathematical expectation values of decreases on them were $\Delta m_1 = 0.01$, $\Delta m_2 = 0.005$ for the relative decrease rates $d\mu_1 = 0.027$, $d\mu_2 = 0.020$. For $Y_E(T_0 n)$ and $X_{EG}(T_0 n)$, the mathematical expectations on these intervals were $m^\circ_{E1} = 0.3505$, $m^\circ_{E1} = 0.3510$, and $m^\circ_{EG1} = 0.490$, $m^\circ_{EG2} = 0.3520$, respectively, on average. The errors of the mathematical expectations estimates were $\Delta m^\circ = 0.0010 - -0.0015$, which is 10–30% from the mathematical expectation values obtained, and this led to successful recognition of decreases with the relative decrease rates of 0.02–0.03.

Testing on experimental MH data made it possible to draw a conclusion about the efficiency of the quasi-Gaussian filtering method and its satisfactory capabilities for recognizing small fluctuations of the mathematical expectations.

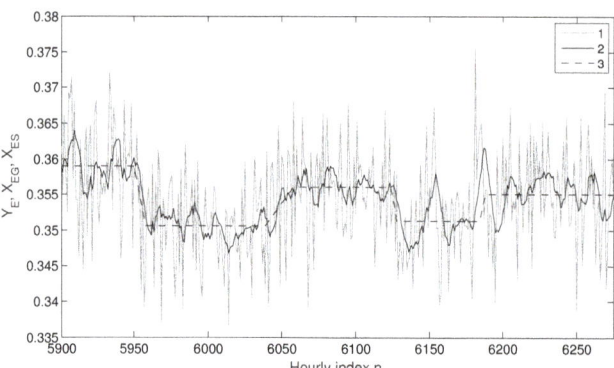

Figure 5. Results of quasi-Gaussian filtering and identification of regions with Forbush decreases: original data (line 1), filtered data (line 2), recognized intervals of various muon flux intensity (line 3).

4. Discussion

The comparison between the model data filtering result obtained using the proposed filter and the one obtained using the fir1 (plots on Figure 3) shows that the resulting time series are close to each other; however, the $X_{MG}(T_0 i)$ seems to be closer to the initial model. The main quantitative result of testing the method and the quasi-Gaussian filter on model normalized Poisson datasets included the calculations for (14) for a set of realizations/ The resulting errors σ_{MG} for $X_{MG}(T_0 i)$, on average, by 15–30% less errors σ_{MF} for $X_{MF}(T_0 i)$. This means that the proposed filtering method provided better filtering (noise reduction) than the standard FIR filter. Consideration of the results of statistical modeling made it possible to draw a conclusion about the efficiency of the method and the quasi-Gaussian filter.

Further tests of the new method on model data, aimed at estimating the mathematical expectation and its RMS errors with respect to the durations and magnitudes of model decreases, showed the method capabilities in disturbance recognitions. It can be seen on Figure 4 that the coefficients ε_{gm}° turned out to be less than the values of the coefficients ε_{fm}°, on average, by about 10–30%. The nature of the plots of coefficients $\varepsilon_{g\sigma}^\circ$ and $\varepsilon_{f\sigma}^\circ$ for the RMS for the same parameters $dn_a, d\mu$ is almost the same—the coefficients $\varepsilon_{g\sigma}^\circ$ are less than the values of the coefficients $\varepsilon_{f\sigma}^\circ$, on average, also by ≈10–30%. From the point of view of recognizing decreases in duration and magnitude, the use of a quasi-Gaussian filter is more preferable than a common FIR filter.

Finally, tests made on real experimental datasets from a muon hodoscope display the method application to data processing and recognition of intervals of decreasing and recovering muon flux intensity. Due to the noise reduction in $X_{EG}(T_0 n)$, it became possible to clearly see the intervals of quiet data (Figure 5), intervals with decreases and recoveries and intervals with declines in mathematical expectation; all these recognized intervals were denoted by a line $X_{ES}(T_0 n)$ (index 3 on Figure 5). On the intervals with the boundary points 5900–5954, 6057–6121, 6197–6276 there were quiet data, on the time intervals 5969–6043, 6127–6189 a decrease in mathematical expectation was observed, the time intervals 5955–5970, 6044–6056, 6122–6126, 6190–6196 corresponded to data with decreases and recoveries. On the intervals 5969–6043, 6127–6189, it is quite possible to recognize relative reductions in mathematical expectation. The errors of the mathematical expectations estimates were Δm° = 0.0010–0.0015, which is 10–30% from the mathematical expectation values obtained, and this led to successful recognition of decreases with the relative decrease rates of 0.02–0.03 and an average duration of *mathrm approx* 10 h.

Testing the proposed method and quasi-Gaussian filter for data variants with indices $j_0 = 31, i_0 = 30$, allowed to obtain results that are almost similar to those depicted on Figure 5); the errors in the estimation of the boundary points of the sections during recognition with depressions amounted to $\delta n \approx$ 2–5 h. Thus, testing on experimental MH data allowed us to make a conclusion about the efficiency of the method and the quasi-Gaussian filter and their satisfactory capabilities for recognizing mathematical expectation small in duration and magnitude.

5. Conclusions

The proposed filtering method for time series of normalized Poisson-distributed data, which was based on the developed digital low-pass quasi-Gaussian filter with a finite impulse response, a gain equal to one at low frequencies and non-negative weighting coefficients, turned out to be efficient; the FR of the low-frequency quasi-Gaussian filter of small dimension was characterized by a better approximation to the prototype filter FR than the FR of common FIR filters.

Testing the filtering method based on the quasi-Gaussian filter for the problems of recognizing small in duration and magnitude fluctuation decreases (increases) in mathematical expectations using statistical modeling and statistical tests have confirmed its effectiveness:

- The proposed method provided a decrease in errors in the filtered time series in comparison with the error values for standard FIR filters, by $\approx$15–30%; the method made it possible to recognize the mathematical expectation fluctuations with a relative decrease of 0.02 and duration of $\approx$12–24 h;
- The proposed method and the developed quasi-Gaussian filter provided relative error coefficients for mathematical expectation and root mean square values that appeared to be $\approx$10–30% less than the error coefficients for common FIR filters.

Testing the method and the low-frequency quasi-Gaussian filter on experimental Poisson data made it possible to draw a conclusion about its satisfactory capabilities for recognizing decreases with relative decrease coefficients $\approx$0.020–0.030.

The proposed method of noise reduction and a quasi-Gaussian filter have favorable prospects of using radiation particle counters for digital information processing in problems of experimental physics and measuring technology.

Author Contributions: Conceptualization, V.G., A.S. and I.Y.; methodology, V.G.; software, V.G. and V.C.; validation, M.D., A.G. and A.S.; formal analysis, V.C.; investigation, A.S.; resources, A.G.; data curation, A.D., A.K. and N.O.; writing—original draft preparation, V.G.; writing—review and editing, R.S. and M.D.; visualization, V.G.; supervision, I.Y.; project administration, A.S.; funding acquisition, A.G. All authors have read and agreed to the published version of the manuscript.

Funding: This work was funded by the Russian Science Foundation (project No. 17-17-01215).

Data Availability Statement: Data sharing is not applicable to this article.

Acknowledgments: The results of experiments presented in this research rely on data collected by the Scientific & Educational Centre NEVOD, National Research Nuclear University MEPhI. We acknowledge URAGAN muon hodoscope data provided by the NEVOD institution. This work employed facilities and data provided by the Shared Research Facility "Analytical Geomagnetic Data Center" of the Geophysical Center of RAS (http://ckp.gcras.ru/) (accessed on 19 April 2021). We would like to thank two anonymous reviewers whose valuable comments helped to improve the manuscript and properly demonstrate the results of our research.

Conflicts of Interest: The authors declare no conflict of interest.

References

1. Lloyd, E.; Ledermann, W. (Eds.) *Handbook of Applicable Mathematics*, Russian ed.; Finansy I Statistika: Moscow, Russia, 1989; Volume 1, pp. 45–83. (In Russian)
2. Taylor, F.J. *Digital Filters: Principles and Applications with MATLAB*; J. Wiley I& Sons: New York, NY, USA, 2011; pp. 53–70.
3. Filter Design Matlab Toolbox. Available online: http://matlab.exponenta.ru (accessed on 7 March 2021).
4. Sergienko, A.B. *Digital Signal Processing: Textbook*, 3rd ed.; BHV-Peterburg: St. Petersburg, Russia, 2011; pp. 371–440. (In Russian)
5. Getmanov, V.G. Digital Signal Processing, 2nd ed.; National Research Nuclear University MEPhI: Moscow, Russia, 2010; pp. 210–225. (In Russian)

6. Goncharov, G.A.; Zubatkin, O.Y.; Lopatin, P.A. Calculation of a filter with a frequency response close to a Gaussian curve. *Commun. Equip. Dig.* **1978**, *5*, 43–49. (In Russian)
7. Bugrov, V.N.; Bessonova, E.V. Numerical design of Gaussian digital filters. *Electron. Des. Technol.* **2012**, *3*, 40–42. (In Russian)
8. Rockenbach, M.; Dal Lago, A.; Schuch, N.J.; Munakata, K.; Kuwabara, T.; Oliveira, A.G.; Echer, E.; Braga, C.R.; Mendonça, R.R.S.; Kato, C.; et al. Global Muon Detector Network Used for Space Weather Applications. *Space Sci. Rev.* **2014**, *182*, 1–18. [CrossRef]
9. Grupen, C.; Shwartz, B. *Particle Detectors*, 2nd ed.; Cambridge University Press: New York, NY, USA, 2008; pp. 56–84.
10. Yashin, I.I.; Astapov, I.I.; Barbashina, N.S.; Borog, V.V.; Chernov, D.V.; Dmitrieva, A.N.; Kokoulin, R.P.; Kompaniets, K.G.; Mishutina, Y.N.; Petrukhin, A.A.; et al. Real-time data of muon hodoscope URAGAN. *Adv. Space Res.* **2015**, *56*, 2693–2705. [CrossRef]
11. Barbashina, N.S.; Kokoulin, R.P.; Kompaniets, K.G.; Mannocchi, A.; Petrukhin, A.A.; Timashkov, D.A.; Saavedra, O.; Trinchero, G.; Chernov, D.V.; Shutenko, V.V.; et al. The URAGAN wide-aperture large-area muon hodoscope. *Instrum. Exp. Technol.* **2008**, *51*, 180–186. [CrossRef]
12. NEVOD COMPLEX. National Research Nuclear University MEPhI. Available online: http://www.nevod.mephi.ru (accessed on 6 March 2021).
13. Himmelblau, D.M. *Applied Nonlinear Programming*, Russian ed.; Mir: Moscow, Russia, 1975; pp. 157–193. (In Russian)
14. Panteleev, A.V.; Skavinskaya, D.V. *Metaheuristic Algorithms for Global Optimization*; University Book: Moscow, Russia, 2019; pp. 5–29. (In Russian)
15. Ingber, L.; Oliveira, E.H.; Petraglia, A.L.; Petraglia, M.R.; Machado, M.A.S. *Stochastic Global Optimization and its Applications with Fuzzy Adaptive Simulated Annealing*; Springer: Berlin/Heidelberg, Germany, 2012; pp. 33–62.
16. Global Optimization Matlab Toolbox. Available online: http://matlab.exponenta.ru (accessed on 4 March 2021).
17. Mikhailov, G.A.; Voytishek, A.V. *Numerical Statistical Modeling. Monte Carlo Method*; Yurayt Publishing House: Moscow, Russia, 2018; pp. 126–174. (In Russian)
18. Statistic Matlab Toolbox. Available online: http://matlab.exponenta.ru (accessed on 5 March 2021).
19. Sidorov, R.; Soloviev, A.; Gvishiani, A.; Viktor Getmanov, V.; Mandea, M.; Petrukhin, A.; Yashin, I.; Obraztsov, A. A combined analysis of geomagnetic data and cosmic ray secondaries for the September 2017 space weather event studies. *Russ. J. Earth Sci.* **2019**, *19*, ES4001. [CrossRef]
20. Oshchenko, A.A.; Sidorov, R.V.; Soloviev, A.A.; Solovieva, E.N. Overview of anomaly measure application for estimating geomagnetic activity. *Geophys. Res.* **2020**, *21*, 51–69. (In Russian)

Article

The August 2018 Geomagnetic Storm Observed by the High-Energy Particle Detector on Board the CSES-01 Satellite

Francesco Palma [1,*,†], Alessandro Sotgiu [1,2], Alexandra Parmentier [1], Matteo Martucci [1,2], Mirko Piersanti [3], Simona Bartocci [1], Roberto Battiston [4,5], William Jerome Burger [5,6], Donatella Campana [7], Luca Carfora [1,2], Guido Castellini [8], Livio Conti [1,9], Andrea Contin [10,11], Giulia D'Angelo [3], Cinzia De Donato [1], Cristian De Santis [1], Francesco Maria Follega [4,5], Roberto Iuppa [4,5], Ignazio Lazzizzera [4,5], Nadir Marcelli [1,2], Giuseppe Masciantonio [1], Matteo Mergé [1,†], Alberto Oliva [11], Giuseppe Osteria [7], Federico Palmonari [10,11], Beatrice Panico [7], Francesco Perfetto [7], Piergiorgio Picozza [1,2], Michele Pozzato [11], Ester Ricci [4,5], Marco Ricci [12], Sergio Bruno Ricciarini [8], Zouleikha Sahnoun [11], Valentina Scotti [7,13], Roberta Sparvoli [1,2], Vincenzo Vitale [1], Simona Zoffoli [14] and Paolo Zuccon [4,5]

1 INFN-Sezione di Roma Tor Vergata, V. della Ricerca Scientifica 1, I-00133 Rome, Italy; alessandro.sotgiu@roma2.infn.it (A.S.); alexandra.parmentier@roma2.infn.it (A.P.); matteo.martucci@roma2.infn.it (M.M.); simona.bartocci@roma2.infn.it (S.B.); luca.carfora@roma2.infn.it (L.C.); livio.conti@uninettunouniversity.net (L.C.); cinzia.dedonato@roma2.infn.it (C.D.D.); cristian.desantis@roma2.infn.it (C.D.S.); nadir.marcelli@roma2.infn.it (N.M.); giuseppe.masciantonio@roma2.infn.it (G.M.); matteo.merge@roma2.infn.it (M.M.); piergiorgio.picozza@roma2.infn.it (P.P.); roberta.sparvoli@roma2.infn.it (R.S.); vincenzo.vitale@roma2.infn.it (V.V.)
2 Department of Physics, University of Rome "Tor Vergata", V. della Ricerca Scientifica 1, I-00133 Rome, Italy
3 INAF-IAPS, V. del Fosso del Cavaliere 100, I-00133 Rome, Italy; mirko.piersanti@inaf.it (M.P.); giulia.dangelo@inaf.it (G.D.)
4 Department of Physics, University of Trento, V. Sommarive 14, I-38123 Povo, Italy; roberto.battiston@unitn.it (R.B.); francesco.follega@unitn.it (F.M.F.); roberto.iuppa@unitn.it (R.I.); ignazio.lazzizzera@unitn.it (I.L.); ester.ricci@unitn.it (E.R.); paolo.zuccon@unitn.it (P.Z.)
5 INFN-TIFPA, V. Sommarive 14, I-38123 Povo, Italy; william.burger@tifpa.infn.it
6 Centro Fermi, V. Panisperna 89a, I-00184 Rome, Italy
7 INFN-Sezione di Napoli, V. Cintia, I-80126 Naples, Italy; donatella.campana@na.infn.it (D.C.); giuseppe.osteria@na.infn.it (G.O.); beatrice.panico@na.infn.it (B.P.); francesco.perfetto@na.infn.it (F.P.); valentina.scotti@na.infn.it (V.S.)
8 IFAC-CNR, V. Madonna del Piano 10, I-50019 Sesto Fiorentino, Italy; g.castellini@ifac.cnr.it (G.C.); s.ricciarini@ifac.cnr.it (S.B.R.)
9 Department of Engineering, Uninettuno University, C.so V. Emanuele II 39, I-00186 Rome, Italy
10 Department of Physics, University of Bologna, V.le C. Berti Pichat 6/2, I-40127 Bologna, Italy; Andrea.Contin@bo.infn.it (A.C.); federico.palmonari@bo.infn.it (F.P.)
11 INFN-Sezione di Bologna, V.le C. Berti Pichat 6/2, I-40127 Bologna, Italy; alberto.oliva@bo.infn.it (A.O.); michele.pozzato@bo.infn.it (M.P.); zouleikha.sahnoun@bo.infn.it (Z.S.)
12 INFN-LNF, V. E. Fermi 40, I-00044 Frascati, Italy; marco.ricci@lnf.infn.it
13 Department of Physics, University of Naples "Federico II", V. Cintia 21, I-80126 Naples, Italy
14 Italian Space Agency, V. del Politecnico, I-00133 Rome, Italy; simona.zoffoli@asi.it
* Correspondence: francesco.palma@roma2.infn.it
† At ASI Space Science Data Center (SSDC) also, V. del Politecnico, I-00133 Rome, Italy.

Abstract: On 25 August 2018, a G3-class geomagnetic storm reached the Earth's magnetosphere, causing a transient rearrangement of the charged particle environment around the planet, which was detected by the High-Energy Particle Detector (HEPD) on board the China Seismo-Electromagnetic Satellite (CSES-01). We found that the count rates of electrons in the MeV range were characterized by a depletion during the storm's main phase and a clear enhancement during the recovery caused by large substorm activity, with the key role played by auroral processes mapped into the outer belt. A post-storm rate increase was localized at L-shells immediately above ~3 and mostly driven by non-adiabatic local acceleration caused by possible resonant interaction with low-frequency magnetospheric waves.

Keywords: space weather; geomagnetic storms; LEO satellites; particle detectors

1. Introduction

Magnetic storms represent major signatures of variability in the Sun-Earth interaction. Such events appear as magnetic disturbances caused by bursts of radiation and charged particles emitted from the Sun in the form of coronal mass ejections (CMEs), solar flares, co-rotating interaction regions (CIRs), etc. [1,2]. These nonlinear and multiscale processes involve a vast set of plasma regions in the mutually interacting magnetosphere and ionosphere.

The terrestrial magnetosphere is under the permanent action of the solar wind. An increase in the solar wind dynamic pressure and a southward direction of the interplanetary magnetic field (IMF) are considered among the fundamental factors in magnetic storm development [3,4]. Under the solar wind driver, global changes occur in the magnetosphere following two principal dynamic triggers: magnetic reconnection at the dayside magnetopause [5] and viscous-like interactions causing magnetospheric convection [6]. One major consequence is the change in the fluxes of charged particles that constitute the magnetospheric ring current [7]. In cascade, the magnetosphere, which is mapped to the upper ionosphere through a system of field-aligned currents, can exchange momentum, energy, and particles with the latter by means of a variety of interactions [8,9]. For example, the transport of plasma between the plasmasphere and ionosphere is severely impacted by altered geomagnetic activity, leading to convection-driven erosion and refilling of the plasmasphere [10] or depletion due to reduced upward flux from the perturbed ionosphere [11].

Discerning physical phenomena that mark the solar-terrestrial environment is not the sole goal of the investigation of storm phenomena, since currently, geomagnetic storms and substorms can severely impact infrastructures at the ground level and in space, also posing a hazard to human health [12–17].

On 25 August 2018, the China Seismo-Electromagnetic Satellite (CSES-01) encountered the first strong magnetic storm since its launch on 2 February 2018. In this paper, after a sketch of the CSES-01 mission and the High-Energy Particle Detector (HEPD) in Section 2, a description of the major solar and geomagnetic characteristics of the storm is reported (Section 3). The magnetospheric disturbance was strong enough to trigger a response in the HEPD instrument; this is presented on the basis of HEPD trigger rate variations observed in the MeV energy range as a function of time and the McIlwain L-shell parameter (Section 4). Observations of this storm in a lower energy interval—from other particle detectors on board CSES-01—were previously presented in [18]. We discuss our results and draw our conclusions in Sections 5 and 6, respectively.

2. Data and Methods

2.1. CSES-01 Mission and HEPD Detector

The CSES-01 [19] is the first item of a multi-satellite constellation under construction by several missions scheduled for the next few years. The satellite was designed for the observation of variations in particle fluxes, plasma parameters, and the electromagnetic field and waves, induced by both natural and anthropogenic sources in the near-Earth space. One major goal of this Chinese-Italian space mission is to investigate possible correlations between the above-mentioned perturbations and the occurrence of high-magnitude earthquakes. Other fundamental targets are the study of space weather phenomena [18,20] and cosmic ray propagation [21].

The CSES-01 relies on the Chinese three-axis stabilized CAST2000 platform, and it is flying in a Sun-synchronous polar orbit at a ~507 km altitude with a 97° inclination and a five-day revisit time. Nine scientific payloads are present on board the satellite [22–30], among which is the HEPD particle detector, which was designed and built by the Italian Limadou Collaboration. A schematic representation of the apparatus is reported in Figure 1. The HEPD is made up of a silicon tracking system; a trigger system that includes one plastic scintillator layer segmented into six paddles; a range calorimeter comprising a tower of 16 plastic scintillator planes, a matrix of 3 × 3 LYSO (lutetium–yttrium oxyorthosilicate) scintillator crystals, and an anti-coincidence (VETO) system equipped with 5 plastic scintillator planes, out of which 4 are placed at the lateral sides of the apparatus and 1 at the bottom (see [30,31]).

Thanks to this set of subdetectors, the HEPD is optimized to detect electrons in the energy range between 3 and 100 MeV and protons between 30 and 250 MeV, as well as light nuclei. In addition, the apparatus can detect different particle populations (solar, trapped, galactic, etc.) according to the satellite position (defined by the McIlwain L-shell parameter) and detected energy.

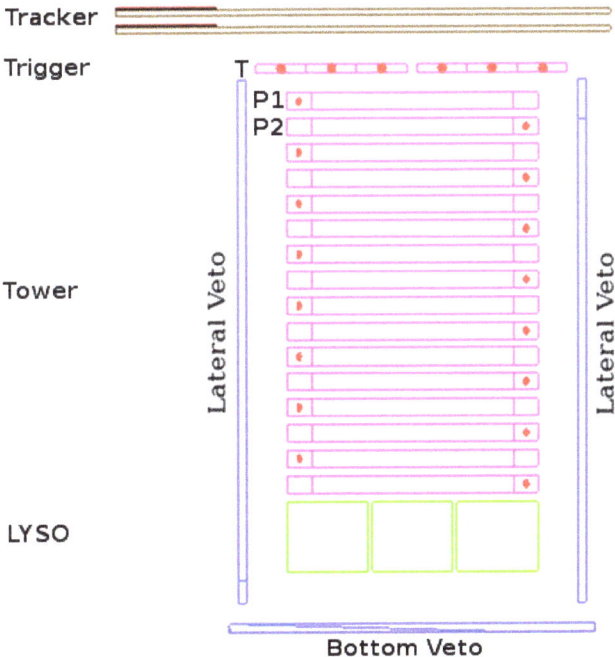

Figure 1. Schematic of the HEPD detector. All mechanical structures (as well as the lateral VETO plane located in the front) have been removed from the figure for visualization purposes.

The transmission of a dedicated command allows setting one of the eight predefined trigger mask configurations [32], which are the result of different logic combinations of counters from the various subdetectors. Hence, the different trigger masks define the aperture and the energy acceptance of the instrument. The trigger condition, labeled as T, corresponds to an above-threshold signal only in the trigger plane, and it is associated with the lowest energy threshold. By requiring a deeper penetration of the particle inside the detector (i.e., using the trigger plane counters and a set of tower planes in "AND" configuration, such as T & P1, T & P1 & P2, and so on), the geometric factor of the HEPD decreases, and consequently, the energy threshold for triggering increases. In July 2018 (late commissioning phase), the HEPD was configured with a trigger condition, labeled as T & P1 & P2, which corresponds to event acquisition and processing only when the released signals in the trigger plane and the first two calorimeter planes (P1, P2) are above predefined thresholds. However, for each of the predefined masks, even when not selected for the online acquisition, a rate meter independently provides the corresponding trigger counting rate (1 s resolution). In this paper, we used the rate meters of three trigger masks (T, T & P1, and T & P1 & P2), corresponding to the integral number of particles per second above different energy thresholds.

Due to adjustments in attitude and additional scheduled maneuvers, the CSES-01 payloads are usually switched off at latitudes below −65° and above +65°. However, the HEPD can benefit from its large field of view (±60°) and geometrical acceptance to collect particles at large L-shells, though for a short time per day. Figure 2 shows the Monte Carlo-based geometrical factor of the HEPD for electrons in three different trigger configurations. In the current one (T & P1 & P2), the geometrical factor reaches a plateau value of ∼500 cm^{2}sr at energies larger than ∼30 MeV.

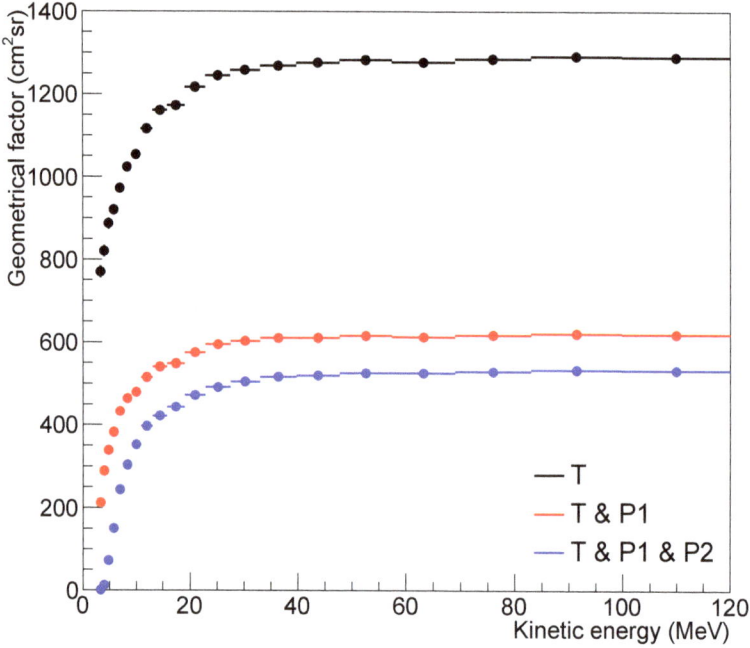

Figure 2. The HEPD geometrical factor for electrons, as estimated from Monte Carlo simulations, as a function of kinetic energy and trigger configuration.

2.2. NOAA19/POES Satellite

NOAA19 is the youngest element in the constellation of the National Oceanic and Atmospheric Administration (NOAA) Polar Orbiting Environmental Satellites (POES), moving along a Sun-synchronous low-Earth orbit (revisit time: 102 min; inclination: 98.7°) at an altitude of ∼850 km, and currently serving as the Prime Service Mission.

The onboard SEM-2 package mounts the Medium Energy Proton and Electron Detector (MEPED) [33], also including two couples of 30°-wide telescopes, of which two are approximately zenith-pointing (MEPED-0°) and two have an azimuthal orientation (MEPED-90°). The electron telescope pair operates in the range from 40 to 2500 keV, over four integral energy channels (E1: > 40 keV; E2: > 130 keV; E3: > 287 keV; E4: > 612 keV).

The Sub-MeV fluxes used in this study were from the MEPED-90° electron telescope in the E2 channel, which represents the best compromise in terms of detection efficiency over the available energy range [34] vs. proton contamination [35].

2.3. DMSP Satellite

The auroral observations used in this study were from the Defense Meteorological Satellite Program (DMSP), consisting of a group of polar, Sun-synchronous satellites flying at ~850 km with a period of ~100 minutes [36]. The DMSP mission focuses on the observation of the near-Earth space plasma environment. In particular, we used the Special Sensor Ultraviolet Spectrographic Imager (SSUSI) instrument on board the DMSP, which is designed to measure far-ultraviolet emissions via imaging spectrograph (SIS) mapping through 5 spectral bins: 121.6 nm (HI Lyman α), 130.4 nm (OI), 135.6 nm (OI), 140–160 nm (N_2 LBHS) and 160-180 nm (N_2 LBHL) [37]. The image resolution is 16 × 156 pixel, while the time resolution (i.e., time to fly above the polar region and acquire an image) is between 20 and 30 min [37].

2.4. RBSP Satellites

The dual-spacecraft Radiation Belt Storm Probes (RBSPs) move along highly elliptical orbits (extending from 1.2 to 5.8 R_E) at an inclination of 10°, thus offering a non-ionospheric point of view due to direct penetration of the radiation belts.

The Relativistic Electron-Proton Telescope (REPT) on board the satellites of the RBSP class measures electrons in differential bins in the energy range ~1–20 MeV with high detection efficiency above 5 MeV. A background partly due to galactic cosmic rays primarily afflicts the REPT measurements in the highest electron channels [38].

2.5. Magnetopause and Plasmapause Position Models

The magnetopause position was obtained by means of the Tsyganenko [39,40] T01 magnetospheric field model. T01 is a semi-empirical model in which the total magnetospheric field of external origin comes from the sum of the Chapman-Ferraro current and contributions from cross-tail, ring, and field-aligned currents. All these contributions are calculated taking into account the solar wind (SW) dynamic pressure, the interplanetary magnetic field (IMF) configuration, and the Dst index. For the present analysis, we used the SW and IMF observations stored in the OMNI CDAWeb repository.

The plasmapause location was assessed using the Liu and Liu [41] model, which is based on the experimental THEMIS-D satellite plasmapause crossing database. The model is based on the following equation:

$$\phi = 2\pi(MLT/24),$$

$$\text{Ł}_{pp} = a1 \cdot [1 + a_{MLT}\cos(\phi - 2\pi a_{phi}/24)] \cdot \log_{10}|Dst_{index}| + \\ + b1 \cdot [1 + b_{MLT}\cos(\phi - 2\pi b_{phi}/24)],$$

where MLT is the magnetic local time. The parameters used in the calculation of $Ł_{pp}$ are reported in Table 1.

Table 1. Parameters used to estimate the plasmapause location.

	a_1	a_{MLT}	a_{phi}	b_1	b_{MLT}	b_{phi}
Dst_{index}	−1.111	−0.2416	21.502	6.013	−0.0565	23.3214

3. The August 2018 Geomagnetic Storm

On 20 August 2018, a large-scale filament gradually erupted from a quiet region of the Sun into an interplanetary CME (ICME) that affected the Earth's environment a few days later, starting on late 25 August 2018 [42] and giving rise to the third largest storm of Solar Cycle 24.

Figure 3 shows the SW parameters as retrieved by the ACE satellite (at the Lagrangian L1 point) from 23–31 August 2018. The magnetic cloud impinged the Earth's magnetosphere between 25 August at ~12:15 UT and 26 August at ~10:00 UT. Looking at Figure 3, we were able to determine the ICME boundaries [43] using the IMF behavior (Panel a) in conjunction with the SW temperature (Panel c) and the SW dynamic pressure (Panel d). Indeed, upon the ICME's arrival, the SW temperature decreased from $\sim 9 \times 10^4$ K to $\sim 1.5 \times 10^4$ K, while the IMF increased to 18 nT, lasting for approximately 12 h. At the same time, the IMF underwent a smooth rotation, leading to a prolonged ($\sim$22 h) southward orientation (Panel b) at $\approx$14:30 UT on 25 August. Finally, the solar wind dynamic pressure fluctuated between $\sim$4 nPa and $\sim$10 nPa. A CIR followed on 26 August: the SW temperature increased from $\sim \times 10^4$ K to nearly 30×10^4 K around 12:20 UT, with the pressure increasing from $\sim$2 nPa to $\sim$8 nPa, as the solar wind stream was crossing a negative polarity high-speed stream (HSS) [42].

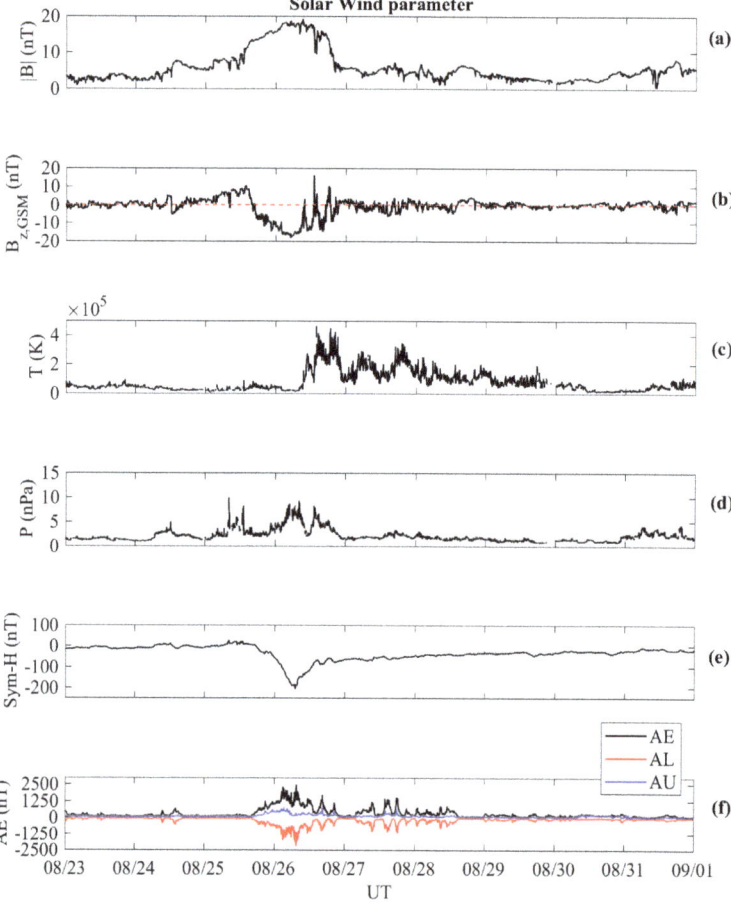

Figure 3. Solar wind parameters observed by the ACE spacecraft at L1: (**a**) IMF intensity; (**b**) IMF B_z component; (**c**) proton temperature; (**d**) dynamic pressure of the solar wind; (**e**) Sym-H index; (**f**) AE (black), AL (red), and AU (blue) indices. The SW parameters are expressed in the Geocentric Solar Magnetospheric (GSM) coordinate system.

To evaluate the consequences of the ICME impact on the Earth's environment, we used the Sym-H index (which mirrors the dynamics of the symmetric part of the ring current [44]) and the AE index (which indirectly measures the energy deposition rate in the polar ionosphere [45]). On 26 August, Sym-H (Figure 3e) showed a rapid decrease, reaching its minimum value (∼ −190 nT) at 07:57 UT. This structure mimics the behavior of the B_z component of the IMF, which shows a long-lasting (∼10 h) negative value starting at 16:52 UT of 25 August. This trend is clearly related to the southward IMF carried by the magnetic cloud [43]. As a consequence of this long interval of negative B_z, the SW plasma could flow inside the Earth's magnetosphere, possibly due to the occurrence of magnetic reconnection at the magnetopause between the geomagnetic field and the IMF [42].

The large bursts in the AE index (black line in Figure 3f) can be related to a sequence of fast relaxation events, possibly stemming from an activity in the near-Earth magnetotail regions in the form of a sequence of loading-unloading releases of energy [46,47]. Such processes give rise to a great amount of particle precipitation in the high-latitude region as confirmed by the behavior of the AL index (red line), which is excellently correlated with the negative turn of B_z. AL (and hence AE) peaks are directly related to the north-south flip of B_z between 11:00 UT and 21:00 UT on 26 August, induced by the arrival of the CIR.

4. HEPD Response to the August 2018 Storm

Figure 4 illustrates a comparison between the HEPD count rate maps before (20–23 August; upper panel) and after the impact of the storm (25–27 August; lower panel). In the top panel, the southern polar region presents a larger trigger rate than the northern one. This is due to the dipole tilt angle, which, in August, allowed the CSES-01 to explore higher geomagnetic latitudes in the Southern Hemisphere than in the Northern Hemisphere. In the bottom panel, an increase in the count rate is evident at both northern and southern latitudes—especially in the southern region—as a consequence of the storm's arrival. Both maps are related to trigger configuration T in Section 2, which requires an above-threshold signal only in the trigger plane and allows detecting the lowest energetic electrons (>3 MeV). As concerns protons, their contribution to the trigger rate increase is negligible due to the absence of direct injection from solar energetic particles (SEPs) during this specific storm event [48]. For visualization purposes, we excluded the South Atlantic Anomaly (SAA) region, which is characterized by extremely high particle rates. For this purpose, we selected magnetic field values larger than 23,000 nT. For this analysis, we calculated magnetic field values by using the International Geomagnetic Reference Field (IGRF) series of mathematical models, in particular the IGRF-12 candidate [49].

The increased particle rate, during the storm time, is also visible as a function of the L-shell and time in Figure 5. The first three panels show the HEPD count rates for three different trigger configurations: from top to bottom, T, T & P1, and T & P1 & P2. The increase in the number of calorimeter planes used for trigger generation resulted in a higher energy threshold for electron detection (>3 MeV, >4.5 MeV, and >8 MeV, respectively), thus reducing the particle rate. For comparison, the time evolution of the Dst index is shown in the bottom panel of Figure 5. As can be inferred by a strong decrease of the Dst down to ∼−190 nT, the start of the storm's main phase was on late 25 August, exactly in coincidence with the increase of the HEPD particle rates.

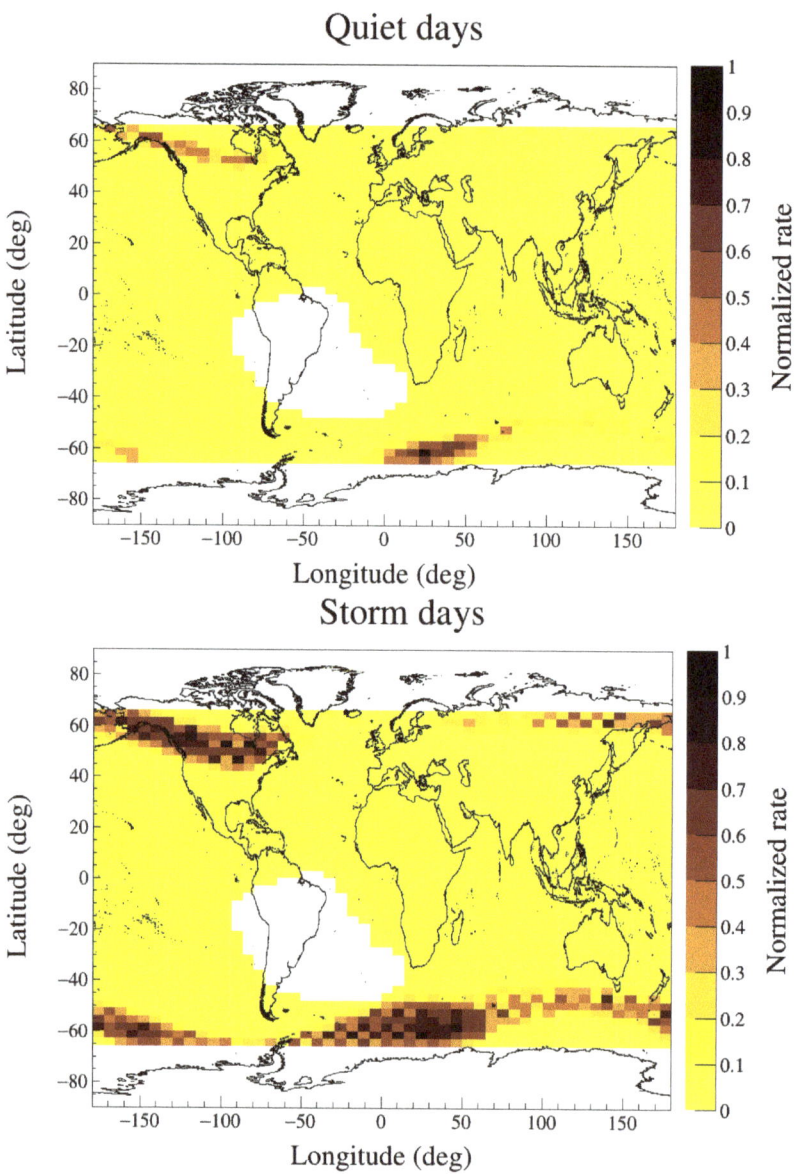

Figure 4. Comparison between an HEPD trigger rate map before the occurrence of the geomagnetic storm, from 20–23 August (upper panel), and after the impact of the storm, from 25 August to 27 (lower panel). The maps are related to trigger configuration T, requiring an above-threshold signal just in the trigger plane and providing the lowest energy threshold for electron detection (>3 MeV). For visualization purposes, we excluded the South Atlantic Anomaly region, characterized by extremely high particle rates.

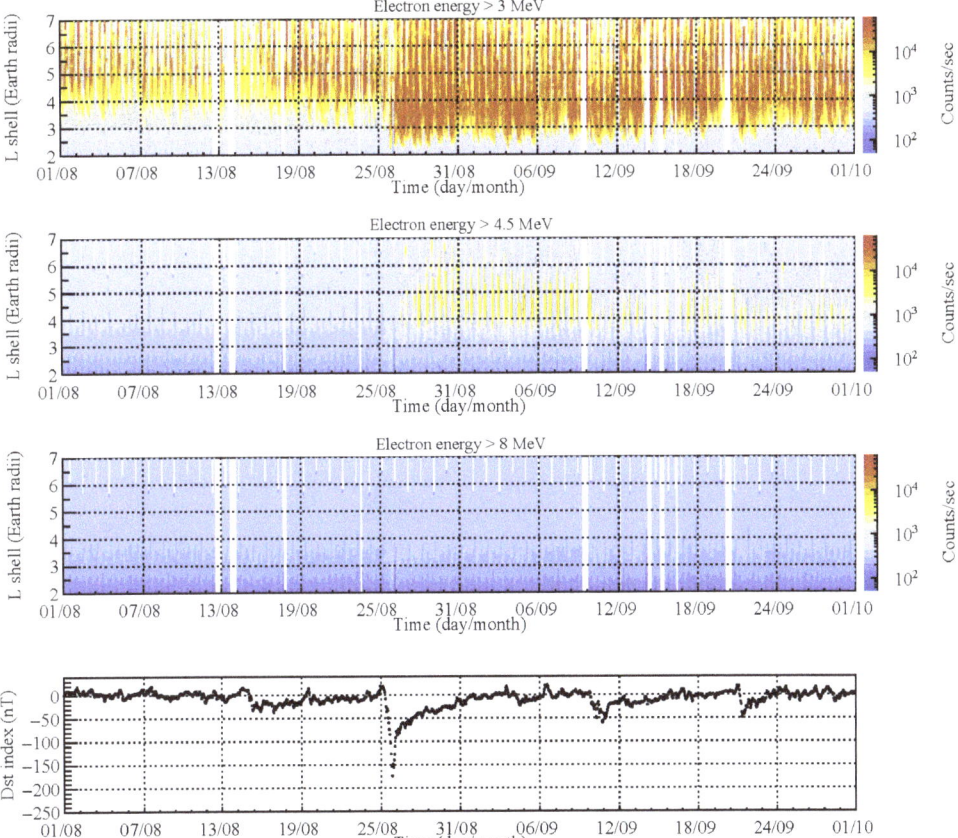

Figure 5. Top three panels: Trigger rates for three different HEPD configurations over the period August-September 2018; from top to bottom, T, T & P1, and T & P1 & P2. Adding more calorimeter planes to the trigger configuration results in increasing the energy threshold for electron detection (>3 MeV for T, >4.5 MeV for T & P1, and >8 MeV for T & P1 & P2). The proton contribution to the trigger rate increase is negligible due to the absence of direct injection from SEPs. The vertical white lines are due to a lack of data. Bottom panel: Time evolution of the Dst index.

5. Discussion

The hit of the ICME gives rise to a compression of the magnetosphere and a backward motion of the plasmasphere, as shown in Figure 6. Indeed, the magnetopause (black line), modeled using the T01 model [39,40] at the moment of the minimum value of the main phase of the geomagnetic storm (i.e., Dst minimum), steps back from $\sim 10 R_E$ before the storm down to $\sim 7.7 R_E$ (R_E being the Earth's radius); while the plasmapause, evaluated by the model of Liu and Liu [41], moves from $\sim 5 R_E$ down to $\sim 3.8 R_E$. On the other hand, the position of the inner boundary of the outer radiation belt (ORB) should reach $L_{ORB} = 3.5$, in accordance with the relation of Tverskaya [50] ($|SymH|_{max} = c \cdot L_{ORB}^{-4}$, where $c = 3 \times 10^4$ nT).

A depletion of the particle count rate (Figure 5) during the main phase of the storm is followed by a clear enhancement during recovery, which coincides with large substorm activity (>1000 nT) as measured by the AL index (red line in Figure 3f). The increase can be spotted at L-shells $\gtrsim 3$ for energies above 3 MeV (Figure 5, top panel) and, to a lesser extent, at L-shells $\gtrsim 4$ for energies above 4.5 MeV (Figure 5, second panel).

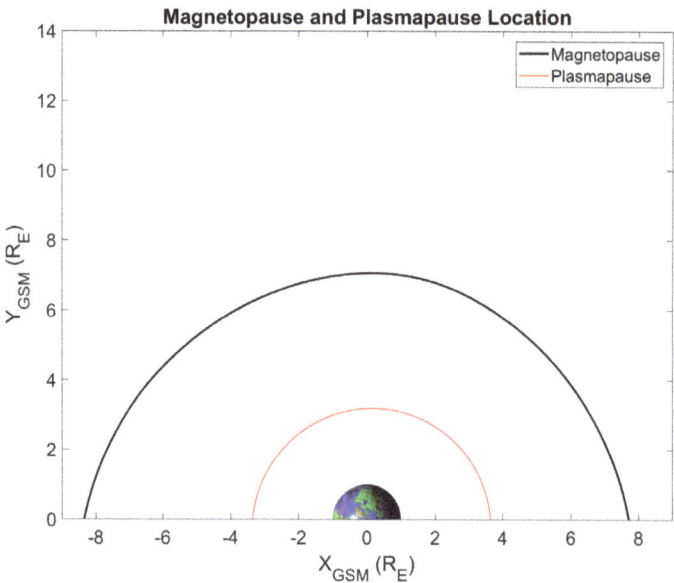

Figure 6. Magnetopause (black line) and plasmapause (red line) profiles as evaluated using the Tsyganenko T01 and the Liu (2014) models, respectively. The reference frame is the GSM. The SW parameters used in the T01 model are: $P = 8.7$ nPa; $B_{y,IMF} = 4.5$ nT; $B_{z,IMF} = -16.8$ nT; Dst $= -174$ nT. The Dst index value used in the plasmapause model is the same as in T01.

As a comparison, the >130 keV electron fluxes measured over the same period by the 90° Medium Energy Proton and Electron Detector (MEPED) on board the NOAA19 satellite are reported in Figure 7. The azimuthal telescopes of the MEPED class are affected by small >280 keV proton contamination even at large L-shells in disturbed periods (<0.6% at 4 < L < 7) [35].

The sub-MeV MEPED-90° observations appear fairly consistent with their MeV counterpart detected by the HEPD, yet reaching flux peaks two orders of magnitude larger than the count rates captured at higher energies. The arrival of the ICME triggered a clear slot-filling event that lasts several days after the impact, with flux enhancements reaching L-shells lower than those occupied by MeV electrons, in accordance with the apparent "barrier" revealed by the Van Allen Probes to significant inward transfer of ultrarelativistic electrons below L $\sim$2.5 [51].

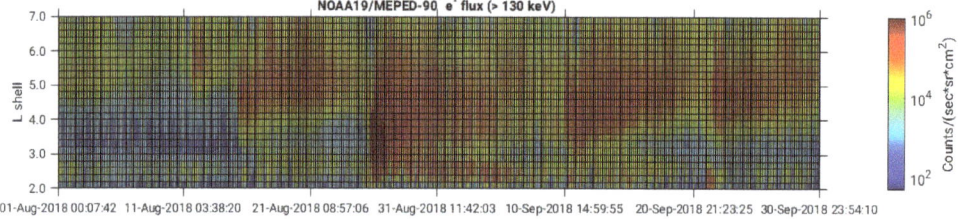

Figure 7. Integral electron fluxes (>130 keV) measured by the MEPED-90° directional telescope on board the NOAA19 satellite over the period Aug-Sept 2018. As reported in [35], proton contamination remains modest in this class of azimuthal detectors even at large L-shells in disturbed periods (<0.6% at 4 < L < 7).

Prolonged and intense substorm activity during recovery (Figure 3f) shows that auroral processes play a non-negligible role in the analysis of particle acceleration in the

ORB and that electrons can undergo a quick acceleration on typical timescales of auroral substorms. Taking into account that the auroral oval (AO) is mapped into the outer portion of the ring current (see [52] and the references therein) can help us better understand the connection between auroral processes and ORB dynamics. Indeed, Figure 4 shows a clear lowering of the AO region in concurrence with the storm (lower panel) with respect to the quiet reference conditions (upper panel). Such results are in agreement with ultraviolet (UV) observations by DMSP/SSUSI [37,53] for both hemispheres (Figure 8c,d). Indeed, UV images in the right column of Figure 8 display clear auroral precipitation at lower latitudes than during pre-storm conditions (left column) as a consequence of the high level of geomagnetic activity induced by south-oriented $B_{z,IMF}$ (see Figure 3) and the expansion of the AO boundaries towards lower latitudes (red dashed lines in Figure 8).

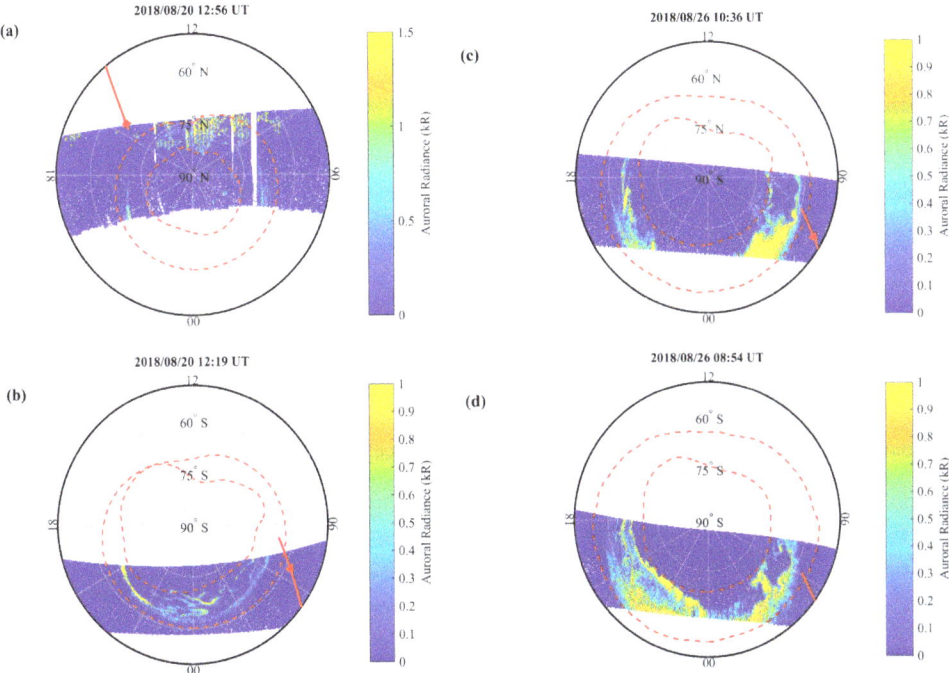

Figure 8. The auroral ultraviolet images by the SSUSI instrument on board the DMSP satellite. Panel (**a**) refers to the quiet Northern Hemisphere observations; Panel (**b**) refers to the quiet Southern Hemisphere observations; Panel (**c**) refers to the stormy Northern Hemisphere observations; Panel (**d**) refers to the stormy Southern Hemisphere observations; the red dashed curves refer to the upper and lower boundaries of the auroral oval. The red curve refers to CSES-01's orbit. The selected reference is the Altitude Adjusted Corrected Geomagnetic (AACGM) coordinate system.

Peaks in the particle count rate during the recovery phase pinpoint a phenomenon of electron acceleration in the radiation belts, which could be ascribed to either adiabatic radial transport or nonadiabatic local heating by resonant interaction with very-low-frequency (VLF) waves [54]. Indeed, in order to make a discrimination between the two drivers, one can observe the radial phase space density (PSD) profiles of REPT electrons—directly monitored in the core of the belts—in the invariant (μ,K,L^*) space at fixed μ and K.

Radial diffusion moves (mostly 90° pitch angle, i.e., equatorially mirroring) electrons across different L^* values while the μ and K invariants remain conserved, thus producing

PSDs with a monotonic decrease from the outer source. Conversely, local acceleration makes PSDs increase over a limited L^* range, with local peaks and negative radial gradients at higher L^* values. Here, we set K at 0.1 $R_E G^{\frac{1}{2}}$ under T04 field modeling, which addressed electrons with pitch angles generally greater than 45° and measured nearly continuously by the REPT instrument over a still broad range of L^* [55]. On the other hand, electron energy corresponding to a particular μ changes with L^*, such that, fixing μ at 4500 $\frac{MeV}{G}$, the range between ~3 MeV and ~7 MeV can be monitored.

Following the evolution of REPT PSDs from late 26 August to early 30 August (Figure 9), at higher L^* values, peak structures with negative gradients were recovered, which, along the previous revelation of persistent chorus waves during the August 2018 event [56], calls for a dominance of local heating.

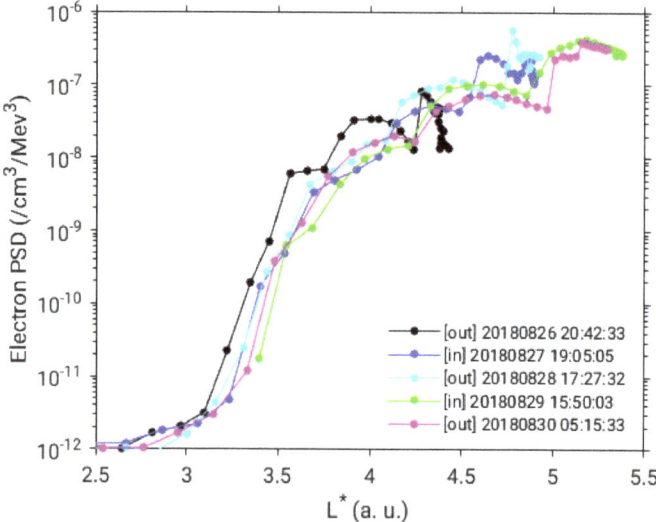

Figure 9. Radial PSD profiles at constant μ (4500 $\frac{MeV}{G}$) and K (0.1 $R_E G^{\frac{1}{2}}$) for RBSP-A/REPT electrons of energies between approximately 3 MeV and 7 MeV along both inbound and outbound crossings of the radiation belts. Complementary RBSP-B/REPT profiles (not shown) are consistent with their A counterpart.

6. Conclusions

The study of geomagnetic storms and other space weather phenomena is crucial to better understand the mechanisms taking place during solar events and to prevent their effects on technological and anthropic systems, such as reduced satellite operations, failures in spacecraft electronics, radio communication problems, etc.

On 25 August 2018, the CSES-01/HEPD particle rate meters were able to detect the effects of a G3-class, ICME-driven geomagnetic disturbance characterized by marked magnetosphere compression and plasmasphere erosion.

In our analysis, a depletion of HEPD count rate at the storm's main phase was observed, followed by a clear rate enhancement during its recovery phase. This increase was detected at L-shells $\gtrsim 3$ for electron energies above 3 MeV, and, to a lesser extent, at L-shells $\gtrsim 4$ for electron energies above 4.5 MeV. These results were consistent with the behavior of integral sub-MeV fluxes measured by the MEPED-90° electron telescope on board the NOAA19/POES satellite, over the same period. The enhancement of HEPD trigger rates suggested a phenomenon of acceleration of energetic electrons, which lasted several days. The discrimination between adiabatic radial transport and non-adiabatic local heating was made by inspection of the PSD profiles of the REPT magnetospheric electrons from late 26 August to early 30 August, whose negative gradients were in favour of the latter, with corroboration by persistent chorus wave interactions previously revealed by other payloads on board the CSES-01 for the same storm.

During the recovery phase, the HEPD trigger rate enhancement was in coincidence with prolonged and intense substorm activity (> 1000 nT), as measured by the AL index. This followed the HEPD orbits crossing the auroral oval region, which expanded during the storm and was marked by UV enhancement especially in the Southern Hemisphere, as detected by the SSUSI instrument on board the DMSP satellite. This occurrence showed that auroral processes cannot be neglected when studying the dynamics of particle acceleration in the ORB, since energetic electrons undergo quick acceleration on typical timescales of auroral substorms.

Considering the sky-rocketing focus on space weather studies in this last decade, HEPD's results prove promising, especially in view of the already-planned constellation of CSES satellites in the next few years (CSES-02 is currently under construction). It is worth noticing that this set of satellites will take shape in a period when several other missions, which contributed to the monitoring of the near-Earth environment, will be either deactivated or well beyond the end of their scheduled lifetimes.

Author Contributions: Writing—original draft, F.P. (Francesco Palma); conceptualization, M.M. (Matteo Martucci), M.P. (Mirko Piersanti), and R.S.; writing—review and editing, A.S., A.P., M.M. (Matteo Martucci), M.P. (Mirko Piersanti), F.M.F., R.I., A.O., E.R., and R.S.; designing the experiment or calibration or data production and processing, A.S., A.P., S.B., R.B., W.J.B., D.C., L.C. (Luca Carfora), G.C., L.C. (Livio Conti), A.C., G.D., C.D.D., C.D.S., F.M.F., R.I., I.L., N.M., G.M., M.M. (Matteo Mergé), A.O., G.O., F.P. (Federico Palmonari), B.P., F.P. (Francesco Perfetto), P.P., M.P. (Michele Pozzato), E.R., M.R., S.B.R., Z.S., V.S., V.V., S.Z., and P.Z. All authors read and agreed to the published version of the manuscript.

Funding: This research received no external funding.

Institutional Review Board Statement: Not applicable.

Informed Consent Statement: Not applicable.

Data Availability Statement: CSES/HEPD data can be found at www.leos.ac.cn/ (accessed on 3 April 2021).

Acknowledgments: This work made use of data from the CSES-01 mission, a project funded by the China National Space Administration (CNSA), the China Earthquake Administration (CEA) in collaboration with the Italian Space Agency (ASI), the National Institute for Nuclear Physics (INFN), the Institute for Applied Physics (IFAC-CNR), and the Institute for Space Astrophysics and Planetology (INAF-IAPS). We kindly acknowledge the OMNIWeb website (https://omniweb.gsfc.nasa.gov/, accessed on 3 April 2021) for providing part of the data used in this paper. This work was supported by the Italian Space Agency in the framework of the "Accordo Attuativo 2020-32.HH.0 Limadou Scienza+" (CUP F19C20000110005) and the ASI-INFN Agreement No.2014-037-R.0, Addendum 2014-037-R-1-2017. M. Piersanti thanks the ISSI-BJ project "the electromagnetic data validation and scientific application research based on CSES satellite" and Dragon 5 cooperation 2020-2024 (ID. 59236).

Conflicts of Interest: The authors declare no conflict of interest.

References

1. Gosling, J.T. The solar flare myth. *J. Geophys. Res.* **1993**, *98*, 18937–18950. [CrossRef]
2. Piersanti, M.; Cesaroni, C.; Spogli, L.; Alberti, T. Does TEC react to a sudden impulse as a whole? The 2015 Saint Patrick's day storm event. *Adv. Space Res.* **2017**, *60*, 1807–1816. [CrossRef]
3. Gonzalez, W.D.; Joselyn, J.A.; Kamide, Y.; Kroehl, H.W.; Rostoker, G.; Tsurutani, B.T.; Vasyliunas, V.M. What is a geomagnetic storm? *J. Geophys. Res.* **1994**, *99*, 5771–5792. [CrossRef]
4. Villante, U.; Piersanti, M. Analysis of geomagnetic sudden impulses at low latitudes. *J. Geophys. Res.* **2009**, *114*, A06209. [CrossRef]
5. Dungey, J. W. Interplanetary magnetic field and the auroral zones. *Phys. Rev. Lett.* **1961**, *6*, 47. [CrossRef]
6. Axford, W.I.; Hines, C.O. A Unifying Theory of High-Latitude Geophysical Phenomena and Geomagnetic Storms. *Can. J. Phys* **1961**, *39*, 1433–1464. [CrossRef]

7. Daglis, I.A. The storm-time ring current. *Space Sci. Rev.* **2001**, *98*, 343–363. [CrossRef]
8. Blanc, M. Magnetosphere-ionosphere coupling. *Comput. Phys. Commun.* **1988**, *49*, 103–118. [CrossRef]
9. Piersanti, M.; Alberti, T.; Bemporad, A.; Berrilli, F.; Bruno, R.; Capparelli, V.; Carbone, V.; Cesaroni, C.; Consolini, G.; Cristaldi, A.; et al. Comprehensive analysis of the geoeffective solar event of 21 June 2015: Effects on the magnetosphere, plasmasphere, and ionosphere systems. *Sol. Phys.* **2017**, *292*, 169. [CrossRef]
10. Pezzopane, M.; Del Corpo, A.; Piersanti, M.; Cesaroni, C.; Pignalberi, A.; Di Matteo, S.; Spogli, L.; Vellante, M.; Heilig, B. On some features characterizing the plasmasphere–magnetosphere–ionosphere system during the geomagnetic storm of 27 May 2017. *Earth Planets Space* **2019**, *71*, 77. [CrossRef] [PubMed]
11. Wang, C.; Zhang, Q.; Chi, P.J.; Li, C. Simultaneous observations of plasmaspheric and ionospheric variations during magnetic storms in 2011: First result from Chinese Meridian Project. *J. Geophys. Res. Space Phys.* **2013**, *118*, 99–104. [CrossRef]
12. Baker D.N. Satellite Anomalies due to Space Storms. In *Space Storms and Space Weather Hazards, NATO Science Series, Series II: Mathematics, Physics and Chemistry*; Daglis, I.A., Ed.; Springer: Dordrecht, The Netherlands, 2001; Volume 48.
13. Ginet, G.P. Space Weather: An Air Force Research Laboratory Perspective. In *Space Storms and Space Weather Hazards, NATO Science Series, Series II: Mathematics, Physics and Chemistry*; Daglis, I.A., Ed.; Springer: Dordrecht, The Netherlands, 2001; pp. 437–457.
14. Kappenman, J.G. An Introduction to Power Grid Impacts and Vulnerabilities from Space Weather. In *Space Storms and Space Weather Hazards, NATO Science Series, Series II: Mathematics, Physics and Chemistry*; Daglis, I.A., Ed.; Springer: Dordrecht, The Netherlands, 2001; Volume 38.
15. Lanzerotti, L.J. Space Weather Effects on Communications. In *Space Storms and Space Weather Hazards, NATO Science Series, Series II: Mathematics, Physics and Chemistry*; Daglis, I.A., Ed.; Springer: Dordrecht, The Netherlands, 2001; Volume 38.
16. Pulkkinen, A.; Bernabeu, E.; Thomson, A.; Viljanen, A.; Pirjola, R.; Boteler, D.; Eichner, J.; Cilliers, P.J.; Welling, D.; Savani, N.P.; et al. Geomagnetically induced currents: Science, engineering, and applications readiness. *Space Weather* **2017**, *15*, 828–856. [CrossRef]
17. Hapgood, M. The Great Storm of May 1921: An Exemplar of a Dangerous Space Weather Event. *Space Weather* **2019**, *17*, 950–975. [CrossRef]
18. Yang, Y.Y.; Zhima, Z.R.; Shen, X.H.; Chu, W.; Huang, J.P.; Wang, Q.; Yan, R.; Xu, S.; Lu, H.-X.; Liu, D.-P. The First Intense Storm Event Recorded by the China Seismo-Electromagnetic Satellite. *Space Weather* **2020**, *18*, e2019SW002243. [CrossRef]
19. Shen, X.; Zhang, X.; Yuan, S.; Wang, L.; Cao, J.; Huang, J.; Zhu, X.; Piergiorgio, P.; Dai, J. The state-of-the-art of the China Seismo-Electromagnetic Satellite mission. *Sci. China Technol. Sci.* **2018**, *61*, 634–642. [CrossRef]
20. Zhima, Z.; Hu, Y.; Shen, X.; Chu, W.; Piersanti, M.; Parmentier, A.; Zhang, Z.; Wang, Q.; Huang, J.; Zhao, S.; et al. Storm-Time Features of the Ionospheric ELF/VLF Waves and Energetic Electron Fluxes Revealed by the China Seismo-Electromagnetic Satellite. *Appl. Sci.* **2021**, *11*, 2617. [CrossRef]
21. Bartocci, S.; Battiston, R.; Burger, W.J.; Campana, D.; Carfora, L.; Castellini, G.; Conti, L.; Contin, A.; De Donato, C.; De Persio, F.; et al. Galactic Cosmic-Ray Hydrogen Spectra in the 40–250 MeV Range Measured by the High-energy Particle Detector (HEPD) on board the CSES-01 Satellite between 2018 and 2020. *Astrophys. J.* **2020**, *901*, 8. [CrossRef]
22. Cheng, B.; Zhou, B.; Magnes, W.; Lammegger, R.; Pollinger, A. High precision magnetometer for geomagnetic exploration onboard of the China Seismo-Electromagnetic Satellite. *Sci. China Technol. Sci.* **2018**, *61*, 659. [CrossRef]
23. Cao, J.; Zeng, L.; Zhan, F.; Wang, Z.; Wang, Y.; Chen, Y.; Meng, Q.; Ji, Z.; Wang, P.; Liu, Z.; et al. The electromagnetic wave experiment for CSES mission: Search coil magnetometer. *Sci. China Technol. Sci.* **2018**, *61*, 653. [CrossRef]
24. Diego, P.; Huang, J.; Piersanti, M.; Badoni, D.; Zeren, Z.; Yan, R.; Rebustini, G.; Ammendola, R.; Candidi, M.; Guan, Y.-B.; et al. The Electric Field Detector on Board the China Seismo Electromagnetic Satellite—In-Orbit Results and Validation. *Instruments* **2021**, *5*, 1. [CrossRef]
25. Yan, R.; Guan, Y.; Shen, X.; Huang, J.; Zhang, X.; Liu, C.; Liu, D. The Langmuir Probe onboard CSES: Data inversion analysis method and first results. *Earth Planet. Phys.* **2018**, *2*, 479–488. [CrossRef]
26. Chen, L.; Ou, M.; Yuan, Y.; Sun, F.; Yu, X.; Zhen, W. Preliminary observation results of the Coherent Beacon System onboard the China Seismo-Electromagnetic Satellite-1. *Earth Planet. Phys.* **2018**, *2*, 505–514. [CrossRef]
27. Liu, C.; Guan, Y.; Zheng, X.; Zhang, A.; Piero, D.; Sun, Y. The technology of space plasma in-situ measurement on the China Seismo-Electromagnetic Satellite. *Sci. China Ser. E Technol. Sci.* **2019**, *62*, 829–838. [CrossRef]
28. Lin, J.; Shen, X.; Hu, L.; Wang, L.; Zhu, F. CSES GNSS ionospheric inversion technique, validation and error analysis. *Sci. China Technol. Sci.* **2018**, *61*, 669. [CrossRef]
29. Li, X.Q.; Xu, Y.B.; An, Z.H.; Liang, X.H.; Wang, P.; Zhao, X.Y.; Wang, H.Y.; Lu, H.; Ma, Y.Q.; Shen, X.H.; et al. The high-energy particle package onboard CSES. *Radiat. Detect. Technol. Methods* **2019**, *3*, 22. [CrossRef]
30. Picozza, P.; Battiston, R.; Ambrosi, G.; Bartocci, S.; Basara, L.; Burger, W.J.; Campana, D.; Carfora, L.; Casolino, M.; Castellini, G.; et al. Scientific Goals and In-orbit Performance of the High-energy Particle Detector on Board the CSES. *Astrophys. J. Suppl. Ser.* **2019**, *243*, 16. [CrossRef]
31. Ambrosi, G.; Bartocci, S.; Basara, L.; Battiston, R.; Burger, W.; Campana, D.; Carfora, L.; Castellini, G.; Cipollone, P.; Conti, L.; et al. Beam test calibrations of the HEPD detector on board the China Seismo-Electromagnetic Satellite. *Nucl. Instruments Methods Phys. Res. Sect. A Accel. Spectrometers Detect. Assoc. Equip.* **2020**, *974*, 164170. [CrossRef]

32. Sotgiu, A.; De Donato, C.; Fornaro, C.; Tassa, S.; Scannavini, M.; Iannaccio, D.; Ambrosi, G.; Bartocci, S.; Basara, L.; Battiston, R.; et al. Control and data acquisition software of the high-energy particle detector on board the China Seismo-Electromagnetic Satellite space mission. *Softw Pr. Exp.* **2021**, *51*, 1459–1480. [CrossRef]
33. Evans, D.S.; Greer, M.S. Polar Orbiting Environmental Satellite Space Environment Monitor-2 Instrument Descriptions and Archive Data Documentation. 2004. https://ngdc.noaa.gov/stp/satellite/poes/docs/SEM2Archive.pdf (accessed on 1 June 2021).
34. Asikainen, T., Mursula, K. Correcting the NOAA/MEPED energetic electron fluxes for detector efficiency and proton contamination. *J. Geophys. Res. Space Phys.* **2013**, *118*, 6500–6510. [CrossRef]
35. Rodger, C.J.; Clilverd, M.A.; Green, J.C.; Lam, M.M. Use of POES SEM-2 Observations to Examine Radiation Belt Dynamics and Energetic Electron Precipitation into the Atmosphere. *J. Geophys. Res.* **2010**, *115*, A04102. [CrossRef]
36. Hardy, D.A.; Schmitt, L.K.; Gussenhoven, M.S.; Marshall, F.J.; Yeh, H.C.; Shumaker, T.L.; Hube, A.; Pantazis, J. *Precipitating Electron and ion Detectors (SSJ/4) for the Block 5D/Flights 6–10 DMSP Satellites: Calibration and Data Presentation*; AFGL-TR-84-0317, ADA 157080; Air Force Geophysics Laboratory—Space Physics Division: Hanscom AFB, MA, USA, 1984.
37. Paxton, L.J.; Schaefer, R.K.; Zhang, Y.; Kil, H. Far ultraviolet instrument technology. *J. Geophys. Res. Space Phys.* **2017**, *122*, 2706–2733. [CrossRef]
38. Baker, D.N.; Kanekal, S.G.; Hoxie, V.C.; Batiste, S.; Bolton, M.; Li, X.; Elkington, S.R.; Monk, S.; Reukauf, R.; Steg, S.; et al. The Relativistic Electron-Proton Telescope (REPT) Instrument on Board the Radiation Belt Storm Probes (RBSP) Spacecraft: Characterization of Earth's Radiation Belt High-Energy Particle Populations. *Space Sci. Rev.* **2013**, *179*, 337–381. [CrossRef]
39. Tsyganenko, N.A. A model of the magnetosphere with a dawn-dusk asymmetry, 1, Mathematical structure. *J. Geophys. Res.* **2002**, *107*, SMP 12. [CrossRef]
40. Tsyganenko, N.A. A model of the near magnetosphere with a dawn-dusk asymmetry, 2, Parameterization and fitting to observations. *J. Geophys. Res.* **2002**, *107*, SMP-12. [CrossRef]
41. Liu, X.; Liu, W. A new plasmapause location model based on THEMIS observations. *Sci. China Earth Sci.* **2014**, *57*, 2552–2557. [CrossRef]
42. Piersanti, M.; De Michelis, P.; Del Moro, D.; Tozzi, R.; Pezzopane, M.; Consolini, G.; Marcucci, M.F.; Laurenza, M.; Di Matteo, S.; Pignalberi, A.; et al. From the Sun to Earth: Effects of the 25 August 2018 geomagnetic storm. *Ann. Geophys.* **2020**, *38*, 703–724. [CrossRef]
43. Burlaga, L.; Sittler, E.; Mariani, F.; Schwenn, R. Magnetic loop behind an interplanetary shock: Voyager, helios, and imp 8 observations. *J. Geophys. Res. Space Phys.* **1981**, *86*, 6673. [CrossRef]
44. Wanliss, J.A.; Showalter, K.M. High-resolution global storm index: Dst versus sym-h. *J. Geophys. Res. Space Phys.* **2006**, *111*, A02202. [CrossRef]
45. Ahn, B.-H.; Akasofu, S.-I.; Kamide, Y. The joule heat production rate and the particle energy injection rate as a function of the geomagnetic indices AE and AL. *J. Geophys. Res. Space Phys.* **1983**, *88*, 6275. [CrossRef]
46. Kamide, Y.; Kokubun, S. Two-component auroral electrojet: Importance for substorm studies. *J. Geophys. Res.* **1996**, *101*, 13027–13046. [CrossRef]
47. Consolini, G.; De Michelis, P. Local intermittency measure analysis of AE index: The directly driven and unloading component. *Geophys. Res. Lett.* **2005**, *32*, L05101. [CrossRef]
48. Abunin, A.A.; Abunina, M.A.; Belov, A.V.; Chertok, I.M. Peculiar Solar Sources and Geospace Disturbances on 20–26 August 2018. *Sol. Phys.* **2020**, *295*, 7. [CrossRef]
49. Thébault, E.; Finlay, C.C.; Beggan, C.D.; Alken, P.; Aubert, J.; Barrois, O.; Bertrand, F.; Bondar, T.; Boness, A.; Brocco, L.; et al. International Geomagnetic Reference Field: The 12th generation. *Earth Planets Space* **2015**, *67*, 79. [CrossRef]
50. Tverskaya, L.V. The boundary of electron injection into the earth magnetosphere. *Geomagn Aeron* **1986**, *26*, 864–865.
51. Baker, D.N.; Jaynes, A.; Hoxie, V.C.; Thorne, R.M.; Foster, J.; Li, X.; Fennell, J.F.; Wygant, J.R.; Kanekal, S.G.; Erickson, P.; et al. An impenetrable barrier to ultrarelativistic electrons in the Van Allen radiation belts. *Nature* **2010**, *515*, 531–534. [CrossRef] [PubMed]
52. Antonova, E.E.; Stepanova, M.V.; Moya, P.S.; Pinto, V.A.; Vovchenko, V.V.; Ovchinnikov, I.L.; Sotnikov, N.V. Processes in auroral oval and outer electron radiation belt. *Earth Planets Space* **2018**, *70*, 127. [CrossRef]
53. Paxton, L.J.; Meng, C.I.; Fountain, G.H.; Ogorzalek, B.S.; Darlington, E.H.; Gary, S.A.; Goldsten, J.O.; Kusnierkiewicz, D.Y.; Lee, S.C.; Linstrom, L.A.; et al. SSUSI: Horizon-to-horizon and limb viewing spectrographic imager for remote sensing of environmental parameters. *Ultrav. Technol. IV* **1993** *1764*, 161–176. [CrossRef]
54. Reeves, G.D.; Baker, D.N.; Belian, R.D.; Blake, J.B.; Cayton, T.E.; Fennell, J.F.; Friedel, R.H.W.; Meier, M.M.; Selesnick, R.S.; Spence, H. The global response of relativistic radiation belt electrons to the January 1997 magnetic cloud. *Geophys. Res. Lett.* **1998**, *25*, 3265–3268. [CrossRef]
55. Reeves, G.D.; Spence, H.; Henderson, M.; Morley, S.; Friedel, R.H.W.; Funsten, H.; Baker, D.N.; Kanekal, S.G.; Blake, J.B.; Fennell, J.F.; et al. Electron Acceleration in the Heart of the Van Allen Radiation Belts. *Science* **2013**, *341*, 991. [CrossRef]
56. Zhang, Z.; Chen, L.; Liu, S.; Xiong, Y.; Li, X.; Wang, Y.; Chu, W.; Zeren, Z.; Shen, X. Chorus acceleration of relativistic electrons in extremely low L shell during geomagnetic storm of August 2018. *Geophys. Res. Lett.* **2020**, *47*, e2019GL086226. [CrossRef]

Article

Investigations of Muon Flux Variations Detected Using Veto Detectors of the Digital Gamma-rays Spectrometer

Krzysztof Gorzkiewicz [1,*], Jerzy W. Mietelski [1], Zbigniew Ustrnul [2], Piotr Homola [1], Renata Kierepko [1], Ewa Nalichowska [1] and Kamil Brudecki [1]

[1] Institute of Nuclear Physics Polish Academy of Sciences, Radzikowskiego 152, 31-342 Krakow, Poland; jerzy.mietelski@ifj.edu.pl (J.W.M.); piotr.homola@ifj.edu.pl (P.H.); renata.kierepko@ifj.edu.pl (R.K.); ewa.nalichowska@ifj.edu.pl (E.N.); kamil.brudecki@ifj.edu.pl (K.B.)
[2] Department of Climatology, Jagiellonian University in Krakow, Gronostajowa 7, 30-387 Krakow, Poland; zbigniew.ustrnul@uj.edu.pl
* Correspondence: krzysztof.gorzkiewicz@ifj.edu.pl

Abstract: This paper presents the results of cosmic ray muons flux monitoring registered by a digital gamma-ray spectrometer's active shield made of five large plastic scintillators. In traditional, i.e., analogue active shields working in anticoincidence mode with germanium detectors, the generated data are used only as a gating signal and are not stored. However, thanks to digital acquisition applied in designed novel gamma-ray spectrometers enabling offline studies, it has not only become possible to use generated data to reduce the germanium detector background (cosmic rays veto system) but also to initialize long-term monitoring of the muon flux intensity. Furthermore, various analyses methods prove the relevance of the acquired data. Fourier analyses revealed the presence of daily (24 h), near-monthly (27 days) and over bi-monthly (68 days) cycles.

Keywords: digital gamma-rays spectrometer; cosmic veto; active shield; muons; muon flux periodicity

1. Introduction

Low-background gamma-ray spectrometry is commonly used in research studies of materials characterized by trace concentrations of gamma-ray emitting radioisotopes. Hence, it finds applications in various fields of science, such as from neutrino physics to environmental research [1–3]. In such investigations, it is vital to use advanced shielding systems to reduce the background radiation of gamma-ray detectors (mainly HPGe).

One of the main components of background radiation is terrestrial gamma radiation, which spectrometers' passive shield can efficiently reduce. However, passive shield layers and their width must be appropriately selected in order to minimize the impact of internal gamma radiation from traces of radionuclides present in the shield's construction materials and any isotopes produced by interactions of cosmic rays with those materials. In low-background detection systems, contributions from air radioactivity, namely radon and its daughter isotopes, are not negligible [2,4].

Another important source of background radiation in gamma-ray spectrometers are particles of secondary cosmic rays. At sea level, the secondary particles flux consists of hadrons, neutrons, gamma quanta, electrons, muons, nucleons and antinucleons [5]. Of these charged particles, muons are the most abundant, with a mean energy of around 4 GeV. The intensity of the muon flux depends on the zenith angle θ of the incident particle, which at sea level can be expressed as follows (1):

$$I(\theta) = I(0°)cos^{n(p)}(\theta), \qquad (1)$$

where $n(p)$ is the particle momentum-dependent exponent and $n \approx 2$ for muons with energies of a few GeV [6]. This relation explains the necessity of shielding gamma-ray detectors in both vertical and horizontal directions. The most significant contribution to the radiation background is from particles with $0°$ zenith angle (vertical direction) [1].

Secondary cosmic rays passing through the shielding may deposit their energy in the germanium crystal (producing continuous background component) as well as generate neutrons and photons via several processes such as muon-induced hadronic and electromagnetic cascades, muon capture and muon-induced spallation reactions [2,7,8].

The cosmic rays background component can be reduced by using active shields consisting of detectors (plastic scintillators or multiwire Charpak chamber) surrounding a passive shield. In the case of particle detection (in the preset coincidence time window) by the active shield's detector and the germanium detector, the signal from the latter is not stored. This is the principle of the so-called classical (analogue) cosmic ray veto system.

However, the development of digital signal processing systems allowed one to apply digital analyzers (digitizers) as critical components of nuclear spectroscopy electronics and substitute a few discrete electronic devices used in analogue electronics. Such a device provides information about the registration time, energy and pulse shape of each significant signal generated by the detector. Furthermore, using digital analyzer allows all generated data to be stored for later processing (offline), enabling the application of various data exploration techniques. Since 2018, a low-background, gamma-ray spectrometer with an active shielding and digital acquisition system has been operating in the Department of Nuclear Physical Chemistry, at the Institute of Nuclear Physics Polish Academy of Sciences (IFJ PAN), in Krakow, Poland [1].

In this paper, we present the results of analyses of data generated by the spectrometer's active shielding. As mentioned above, these data are used to reduce the germanium detector background. Furthermore, offline data analysis makes it possible to develop a continuous cosmic-ray muons monitoring system. Hence one device, namely a low-background, digital gamma-ray spectrometer, can simultaneously performs experiments from two branches of physics, namely low-background gamma-ray spectrometry and astrophysics.

2. Materials and Methods

The spectrometer is equipped with a Broad Energy Germanium detector BE5030 (Canberra, USA) with relative efficiency of $\geq 48\%$ and a composite passive shield in a cubic shape with an internal layer made of lead cast over 2500 years ago. The active shielding consists of five large, five cm thick plastic scintillation detectors EJ-200 with photo-multipliers ET 9900 (Scionix, Nl). These detectors are mounted outside the passive shield. TOP and BOTTOM detectors are placed horizontally while the latter three—FRONT, SIDE and REAR—are placed vertically. The relative positioning of all spectrometer's detectors is depicted in Figure 1. The detectors' preamplifier signals are transmitted directly to the inputs of a digitizer DT5725 (CAEN, Italy), where data acquisition and signal pre-processing are performed.

The digital analyzer DT5725 allows simultaneous acquisition of data generated by up to eight detectors (at this moment, in our setup six inputs are occupied) with a maximum time resolution of 4 ns. Registered data consist of pulse time registration, height (proportional to deposited particle energy) and shape. Raw data generated by the digitizer are stored in a PC as six files (in *.csv format), which are further processed using purposely written software VETO. Commissioning and optimization processes of the described spectrometer and software development and its properties are discussed in detail in [1].

The data were collected from 1 September 2018 to 30 April 2020. Since the described spectrometer is primarily used to measure low-active gamma-ray emitting samples, the obtained data may be divided into two groups, namely short and long-period data. Short period data consist of data generated by scintillation detectors during single gamma-rays spectrometric measurement, which last up to 6 days. Data acquisition is stopped at the end of measurements in order to replace the sample in the spectrometer's chamber and/or refill the liquid nitrogen dewar. Such breaks last up to 30 min, after which the subsequent gamma-ray spectrometric measurement along with registration of cosmic ray particles by scintillators is restarted. The long-period data consist of all digitizer output files generated from 1 September 2018 to 30 April 2020. During this time, 256 gamma-spectrometric measurements were carried out, and as a result 1280 output files of scintillation detectors were generated (total size 270 GB). In this paper, data generated by scintillators TOP (horizontal) and FRONT (vertical) were used to analyze the long-term modulations of muon flux.

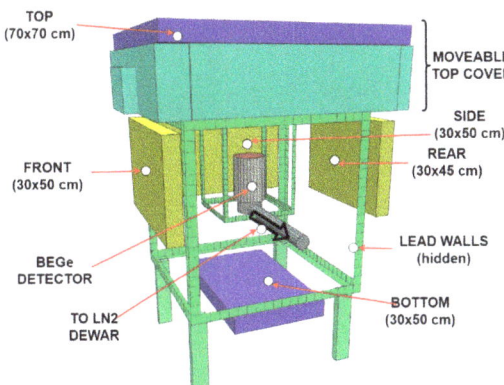

Figure 1. The relative positioning of the low-background, digital gamma-ray spectrometer's detectors. In blue—horizontally (TOP and BOTTOM); in yellow—vertically placed scintillators (FRONT, SIDE and REAR). Each of the scintillators is 5 cm thick. The device is in operation at the Institute of Nuclear Physics Polish Academy of Sciences. Some parts of the passive shield (e.g., the lead walls) are hidden.

For this work, the atmospheric pressure data were obtained thanks to the cooperation with the Institute of Meteorology and Water Management National Research Institute. Data are collected at the Meteorological Station located at the Kraków-Balice Airport. The station is located 6 km west of IFJ PAN, and data were recorded hourly with a high accuracy of 0.1 hPa. The measurements contain pressure values at the station level of 237 m asl (above sea level). It is worth mentioning that the pressure values did not differ from the long-term data series. Only a clear diurnal course was found, which exceeded 1 hPa. The lowest pressure values are recorded in the afternoon, while the highest at night and in the morning. These hours coincide with the strongest and weakest convective movements in the atmosphere, respectively.

In the case of analyses of variations of cosmic rays intensity using detectors located at ground level, atmospheric pressure effects must be taken under consideration [9,10]. The influence of atmospheric pressure on the intensity of the cosmic ray flux is defined as the barometric effect, which includes the following components [11]:

- Absorption having a negative impact on the intensity of the muon flux. With increasing atmospheric pressure (and thus the amount of matter in the air column), the probability of absorption and scattering of particles is increased;
- Decay also demonstrating the negative effect, including the increase in the number of muons decays with increasing atmospheric pressure, which is caused by the increase in the height at which muons are generated;
- Generation defines the positive impact on the muon flux intensity and considers the increase in the number of pions produced with increasing pressure.

Near the Earth's surface, the dominant factor is absorption one; considering that air density is usually the highest near the ground, the knowledge of the atmospheric pressure at the level at which the detection is performed is sufficient for determining the value of the barometric effect [12].

The normalized deviation of the recorded muon flux from the average depends on the change in atmospheric pressure [13]:

$$\frac{\Delta I}{I} = \beta_P \Delta P, \qquad (2)$$

where $\frac{\Delta I}{I}$ is the normalized deviation of muon count rate, ΔP is the deviation of the atmospheric pressure from the mean and β_P is the so-called barometric coefficient.

The β_P coefficient (expressed in (%/hPa)) can be determined by assuming a linear correlation between changes in the normalized muon flux intensity and variations in atmospheric pressure. Therefore, it is essential to estimate the β_P value only during the most geomagnetically quiet days [14].

Pressure corrected data are analyzed using various statistical tools. In order to investigate any correlation between subsequent registered events, the autocorrelation function (ACF) of a sequence of time intervals between pulses generated by detectors is determined. The autocorrelation function indicates the Pearson correlation coefficient between values of the same series as a function of time lag. Furthermore, Fast Fourier Transform (FFT) of a given time series is performed, allowing the investigation of any periodic components. Prior transformation, mean and linear trends have been subtracted from the sequence.

In the case of long-period data, a threshold normalization procedure was necessary. The spectrometer's configuration and optimization procedures allow a fixed threshold level for the scintillators spectra to 300th ADC channel to be set [1]. This procedure allowed redundant data (mainly registered gamma rays) which did not increase the effectiveness of the cosmic ray veto system to be reduced up to around 24 times However, to limit the influence of any signal threshold level fluctuations or gain changes in the digital acquisition circuit which may have occurred in data collected over a long period and cause uncontrolled changes in the recorded number of counts, normalization of the spectra discrimination level was carried out [14]. The normalization procedure involved cutting off part of the energetic spectra located below the ADC channel containing 30% of the maximum number of counts recorded in one channel in a given spectrum (i.e., the highest point in the energy spectrum, see Figure 2). The prepared output files were used to determine a time series of hourly muon count rates for the whole period considered.

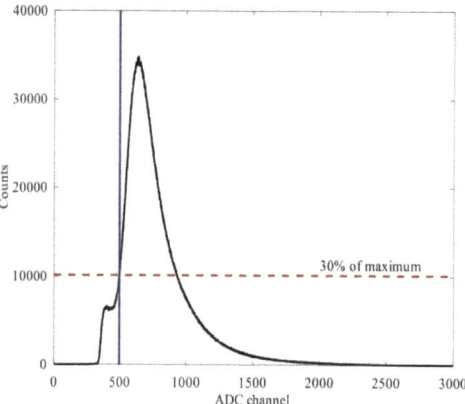

Figure 2. Scintillator energy spectra discrimination procedure to eliminate fluctuations in the threshold level and signal gain. The black curve is the scintillation detector spectrum with a fixed threshold level set to the 300th ADC channel; the red dashed line represents 30% of the maximum number of counts registered in one channel in the given spectrum. The blue vertical line indicates the ADC channel defining the normalized discrimination level. The part of the spectrum below this channel is not used in further analyses.

3. Results and Discussion

3.1. Short Period Data

Distributions of pulses generated by the scintillators in unit time t should correspond to Poisson distributions with an expected value equal to λt. This value (for a chosen time interval) depends mainly on the area of the detector and its vertical or horizontal position. Figure 3 presents distributions of the numbers of counts registered in one second by the scintillation detectors TOP, BOTTOM and FRONT. In order to obtain experimental data (the cyan, red and pink lines in Figure 3), Poisson distributions were fitted with expected count rates λ_{TOP} = 71.93 counts/s (black), λ_{BOTTOM} = 18.88 counts/s (green) and λ_{FRONT} = 14.43 counts/s (blue) for TOP, BOTTOM and FRONT detectors, respectively. The horizontal detectors registered more particles in unit time than the FRONT detector, which was placed vertically. Additionally, Figure 3 demonstrates a great amount of data registered by a digital gamma-ray spectrometry system and supports the demand for the optimization process. Such facts are in accordance with previous research results described in [1]. Further data analyses involved the verification of the presence of correlated structures in detector signals. The gamma-ray measurement investigation lasted approximately 119 h.

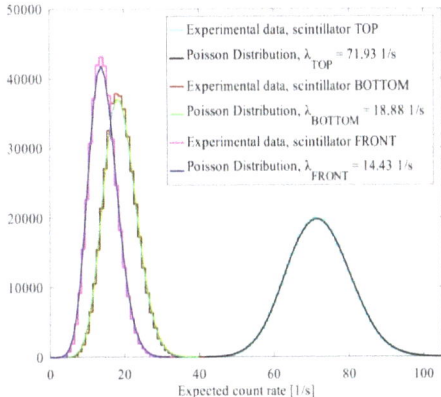

Figure 3. Expected count rate distributions for scintillation detectors: TOP (71.93 counts/s), BOTTOM (18.88 counts/s) and FRONT (14.43 counts/s).

Figure 4a presents values of the ACF in the examined series in cases of up to ten events lag and from 10–1000 events lags. The ACF values oscillate around zero, indicating that there were no correlations in this signal. This means that practically every registered muon can be considered as a single independent event. The same results were achieved for other scintillators, implying that singular scintillators detect non-correlated cosmic-ray muons. Additionally, conducted analyses proved that signals generated by scintillation detectors are stochastic Poisson processes [15].

Figure 4b shows the obtained results from FFT analysis of the sequence of counts registered in one second by the TOP detector. A flat frequency spectrum (and, consequently, flat power spectrum) proves that data generated by the individual scintillation detectors during single gamma-ray spectrometric measurement correspond to white noise, and it is impossible to detect, using single scintillators, correlated muons created in the same air-shower event. This is because the time resolution of an active shield scintillator is mainly determined by two signal shaping parameters: the rise time of the trapezoid generated by the digitizer's energy filter and the length of its flat part [1]. Since the main purpose of the constructed active shield, muon detection, requires correct evaluation of the pulse height, the total shaping time is around 12 μs, much longer than the intervals between any possible registered cosmic-ray particles generated in the same air-showers.

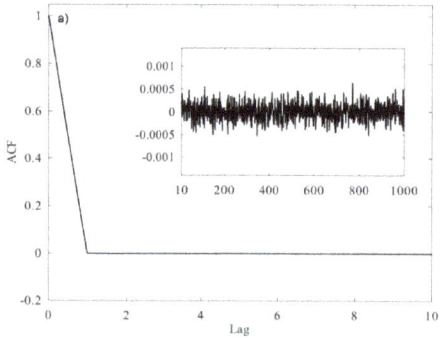 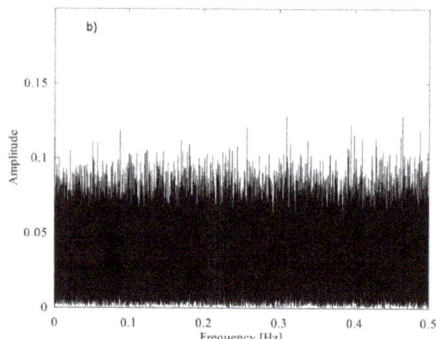

Figure 4. (a) Autocorrelation function of time intervals between events registered at the TOP scintillator. As shown, there is no correlation in the examined series. (b) Discrete Fourier transform analysis of the sequence of counts registered by detector TOP in 1 s. No significant periodicities were found. Similar results were obtained for other detectors: BOTTOM, FRONT, REAR and SIDE.

Bearing in mind the previous results, coincidence events of signals from different scintillators have been studied. Coincidence events from four scintillation detectors, namely TOP, BOTTOM, FRONT and REAR, were found using the VETO software. The detection system registered 471 events of four-fold coincidences during the investigated gamma-spectrometric measurement. Figure 5 presents the distribution of the coincidence events found as a function of the maximum time interval between pulses from the same event. Due to the non-linear geometric positioning of the considered scintillators, generated signals must come from at least two correlated muons from the same air-shower. Furthermore, the width of the distribution (σ = 12 ns) is comparable with the digitizer time resolution, and correlated muons time travel through the distance between scintillators. This feature is interesting from the novel CREDO project point of view [16], which focuses on studying cosmic rays and cosmic-ray ensembles. The presented detection system includes five plastic scintillators, which may play the role of a reference detector, that are complementary to other detection systems (e.g., smartphones' cameras) already applied in the CREDO project [16].

3.2. Long-Period Data

For long-period data, the barometric coefficient β_P was established. During investigations, only data collected on the ten most quiet days of every month were used. The list of those days was acquired from the International Quiet Days (IQD) database (http://wdc.kugi.kyoto-u.ac.jp/qddays/ accessed on 14 February 2021) [17]. Figure 6a presents relative deviation of the muon count rate $\frac{\Delta I}{I}$ as a function of pressure deviation ΔP. The blue markers represent all data from the period studied, while the red markers represent data recorded only on IQD days. The correlation coefficient between these values and the pressure changes indicates strong negative correlation ($r_{1,TOP} = -0.75$). The β_P coefficient was determined by a least squares fit, and its value was $\beta_{P,TOP} = -0.168(2)$ %/hPa. The value was used to correct the data and to eliminate the dependence of the muon flux intensity on the atmospheric pressure using the Equation (2). The obtained results are presented in Figure 6b. The value of the correlation coefficient after data correction $r_{2,TOP} = -0.08$ confirms a significant reduction in this dependence. In the case of the FRONT detector, analyses were performed to allow an estimate of $\beta_{P,FRONT} = -0.153(2)$ %/hPa, which reduced the data correlation from $r_{1,FRONT} = -0.72$ to $r_{2,FRONT} = 0.09$.

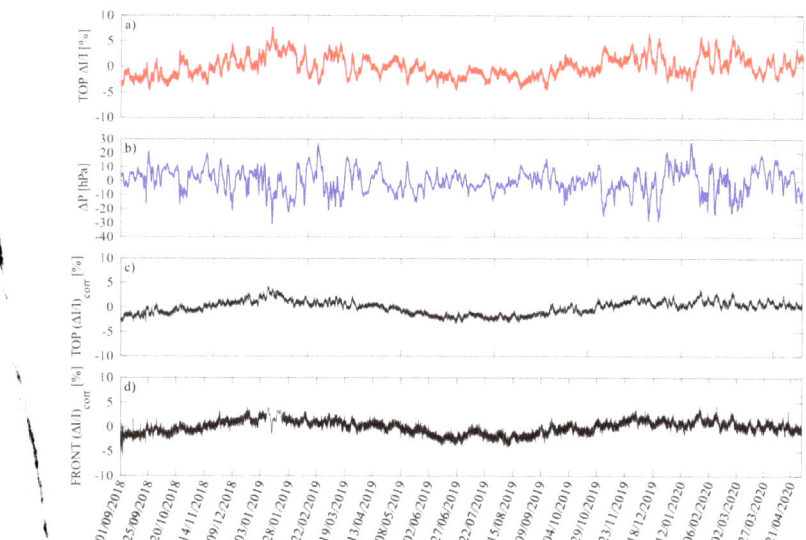

Figure 7. (a) Normalized changes in the muon flux intensity recorded by the TOP detector before correcting for atmospheric pressure, (b) changes in atmospheric pressure, (c) corrected muon flux intensity I_{corr} registered by scintillator TOP and (d) corrected muon flux intensity I_{corr} registered by scintillator FRONT.

Fast Fourier transform analysis was conducted on the pressure-corrected data, and the obtained results are presented in Figure 8.

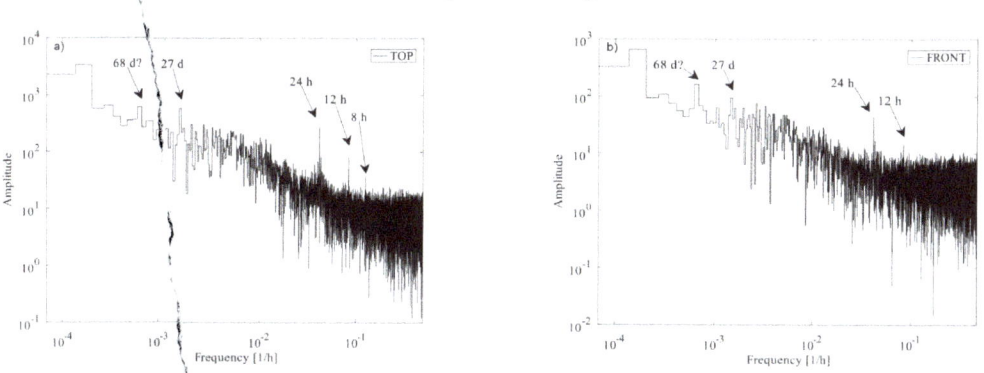

Figure 8. Discrete Fourier analysis of long-term muon intensity registered by scintillation detector TOP (**a**) and FRONT (**b**). Periodicities in time series 24 h (along with harmonics 12 h and 8 h), 27 d and 68 d have been found.

In the case of the TOP detector, the FFT algorithm allowed identification of the diurnal periodicity (24 h) with harmonics components (12 h and 8 h) of muon flux caused by the rotation of the Earth and its relative motion in the solar magnetic field, modulated by solar wind [18,19]. Furthermore, a periodic component with a period of around 27 days was identified. This variability is associated with the effect of the Sun's rotation, causing variations in both intensity of solar wind particles and their speed [18,20]. In addition, a periodicity of about 68 days is noticeable in the frequency spectrum but its origin is unclear. Takai et al. (2016) [21] conducted a frequency analysis of the eight-year time series of muon flux recorded during the MARIACHI experiment. The authors identified a signal component, among others, with a period of about 62.5 days (which is the closest

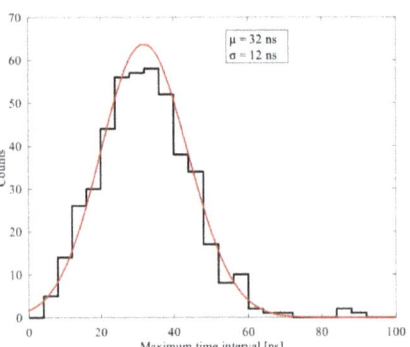

Figure 5. Distribution of 471 coincidence events registered by TOP, BOTTOM, FRONT and REAR detectors. To obtain the data, a Gaussian distribution has been fitted (μ = 32 ns; σ = 12 ns).

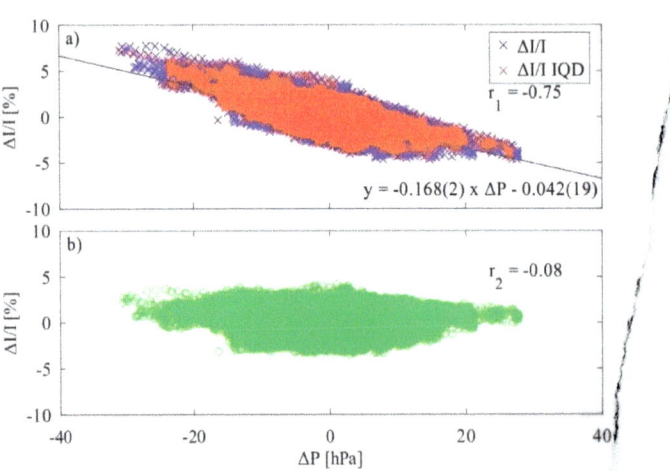

Figure 6. Determination of the barometric coefficient for the data recorded by the TOP detector. (**a**) Relative deviation of muon count rate as a function of pressure deviation for all data (blue markers) and for International Quiet Days (IQD) data—red markers. The correlation coefficient between the IQD data and the pressure variations was $r_1 = -0.75$. Obtained barometric coefficient $\beta_P = -0.168\,\%/\text{hPa}$. (**b**) Pressure corrected data; correlation coefficient reduced to $r_2 = -0.08$.

Figure 7a shows the normalized intensity of the muon flux registered by the detector TOP before correction for changes in atmospheric pressure (which are depicted in Figure 7b). The pressure corrected data demonstrated in Figure 7c,d for scintillator TOP and FRONT, respectively, show less variability. It should be noted that seasonal variation is still clearly visible; however, seasonal changes do not affect the final results for the analyzed period (20 months).

to the result obtained in presented research), but unfortunately its interpretation was not provided. Hence the 68day periodicity needs to be further analyzed.

The results of the FFT analyses of the signal generated by detector FRONT (Figure 8b) confirmed diurnal (with 12 h harmonics). In addition, there are 27 and 68 days periodicities in the horizontal component of cosmic ray flux registered by the vertical scintillator.

4. Conclusions

This paper presents the capabilities of a low-background digital gamma-ray spectrometer which expand its fields of applications and the analysis techniques to correct and extract information of the data generated by an active shield detector. Usually, such data are used only to reduce the germanium detector radiation background; thus, it is not saved or analyzed. Our approach, involving digital data acquisition and offline analysis, allows both tasks to be fulfilled simultaneously.

In the case of short period time series (i.e., data generated by a single scintillator during a single gamma-spectrometric measurement), research studies indicated that, according to expectations, the number of pulses generated by scintillators in unit time (in our case, 1 s) follows a Poisson distribution, and the time intervals between the pulses are not correlated. Moreover, by using the fast Fourier transform, the absence of periodic structures in these series was demonstrated.

Relatively long period data were defined as the hourly mean count rates of the registered cosmic rays. Investigations demonstrated strong negative correlations between the recorded muon flux and atmospheric pressure at ground level. The estimated barometric coefficients allowed data generated by detectors TOP and FRONT to be corrected for pressure. These corrected data were used in the analysis by using the FFT technique, and periodic components have been identified in both time series, including those related to the rotation of the Earth (with a period of 24 h) and the rotation of the Sun (a period of about 27 days). The periodicity with a 68 days period remains unexplained.

To summarize, by using a digital data acquisition system, it is possible to expand the research potential of the low-background gamma-ray spectrometer by numerous methods to explore collected measurement data and to allow monitoring of cosmic-ray muons flux registered by the active shield's detectors. Moreover, this device may find applications in the CREDO scientific project and other investigations focused on various phenomena correlated with the intensity of muon flux (e.g., earthquakes [22]).

Author Contributions: Conceptualization, J.W.M. and K.G.; methodology, K.G., J.W.M. and P.H.; software, K.G.; validation, K.G., J.W.M., Z.U., R.K. and P.H.; formal analysis, K.G.; investigation, K.G., J.W.M., R.K. and Z.U.; data curation, K.G. and Z.U.; writing—original draft preparation, K.G., J.W.M. and R.K.; writing—review and editing, K.G., J.W.M., R.K., Z.U., P.H., E.N. and K.B.; visualization, K.G., E.N. and K.B. All authors have read and agreed to the published version of the manuscript.

Funding: This research received no external funding.

Institutional Review Board Statement: Not applicable.

Informed Consent Statement: Not applicable.

Data Availability Statement: Not applicable.

Conflicts of Interest: The authors declare no conflict of interest.

References

1. Gorzkiewicz, K.; Mietelski, J.W.; Kierepko, R.; Brudecki, K. Low-background, digital gamma-ray spectrometer with BEGe detector and active shield: Commissioning, optimisation and software development. *J. Radioanal. Nucl. Chem.* **2019**, *322*, 1311–1321. [CrossRef]
2. Wen, X.; Zhou, F.; Fukuwa, N.; Zhu, H. A simplified method for impedance and foundation input motion of a foundation supported by pile groups and its application. *Comput. Geotech.* **2015**, *69*, 301–319. [CrossRef]
3. Gilmore, G.; Hemingway, J.D. *Practical Gamma-ray Spectrometry*; Wiley: Chichester, UK, 1995.
4. Núñez-Lagos, R.; Virto, A. Shielding and background reduction. *Appl. Radiat. Isot.* **1996**, *47*, 1011–1021. [CrossRef]

5. Griender, P.K.F. *Cosmic Rays at Earth: Researcher's Reference Manual and Data Book*, 1st ed.; Elsevier Science B.V.: Amsterdam, The Netherlands, 2001; ISBN 0444507108.
6. Cecchini, S.; Spurio, M. Atmospheric muons: Experimental aspects. *Geosci. Instrum. Methods Data Syst.* **2012**, *1*, 185–196. [CrossRef]
7. Jovančević, N.; Krmar, M.; Mrda, D.; Slivka, J.; Bikit, I. Neutron induced background gamma activity in low-level Ge-spectroscopy systems. *Nucl. Instrum. Methods Phys. Res. Sect. A Accel. Spectrom. Detect. Assoc. Equip.* **2010**, *612*, 303–308. [CrossRef]
8. Kudryavtsev, V.A.; Spooner, N.J.C.; McMillan, J.E. Simulations of muon-induced neutron flux at large depths underground. *Nucl. Instrum. Methods Phys. Res. Sect. A Accel. Spectrom. Detect. Assoc. Equip.* **2003**, *505*, 688–698. [CrossRef]
9. Savic, M.; Dragic, A.; Veselinovic, N.; Udovicic, V.; Banjanac, R.; Jokovic, D.; Maletic, D. Effect of pressure and temperature corrections on muon flux variability at ground level and underground. *arXiv* **2016**, arXiv:1701.00164.
10. Maghrabi, A.; Almutayri, M. Atmospheric Effect on Cosmic Ray Muons at High Cut-Off Rigidity Station. *Adv. Astron.* **2016**, *2016*, 1–9. [CrossRef]
11. Dorman, L.I. Cosmic Rays in the Earth's Atmosphere and Underground. 2004. Available online: https://link.springer.com/book/10.1007/978-1-4020-2113-8?page=1#toc (accessed on 24 August 2021). [CrossRef]
12. Dayananda, M.A. Correlation Studies of Cosmic Ray Flux and Atmospheric and Space Weather. Ph.D. Thesis, Georgia State University, Atlanta, GA, USA, 2013.
13. De Mendonça, R.R.S.; Raulin, J.-P.; Echer, E.; Makhmutov, V.; Fernandez, G. Analysis of atmospheric pressure and temperature effects on cosmic ray measurements. *J. Geophys. Res. Space Phys.* **2013**, *118*, 1403–1409. [CrossRef]
14. Savić, M.; Maletic, D.; Joković, D.; Veselinović, N.; Banjanac, R.; Udovičić, V.; Dragic, A. Pressure and temperature effect corrections of atmospheric muon data in the Belgrade cosmic-ray station. *J. Phys. Conf. Ser.* **2015**, *632*. [CrossRef]
15. Ross, S.M. *Introduction to Probability Models*, 10th ed.; 2010; ISBN 978-0-12-375686-2. Available online: http://fac.ksu.edu.sa/sites/default/files/book_solution.pdf (accessed on 24 August 2021).
16. Homola, P.; Beznosko, D.; Bhatta, G.; Bibrzycki, Ł.; Borczyńska, M.; Bratek, Ł.; Budnev, N.; Burakowski, D.; Alvarez-Castillo, D.E.; Almeida Cheminant, K.; et al. Cosmic-Ray Extremely Distributed Observatory. *Symmetry* **2020**, *12*, 1835. [CrossRef]
17. International Q-Days and D-Days. Available online: http://wdc.kugi.kyoto-u.ac.jp/qddays/ (accessed on 14 February 2021).
18. Kudela, K.; Langer, R. On quasi-periodic variations of low-energy cosmic rays observed near earth. *Radiat. Prot. Dosim.* **2015**, *164*, 471–476. [CrossRef] [PubMed]
19. Arunbabu, K.P.; Ahmad, S.; Chandra, A.; Gupta, S.K.; Dugad, S.R.; Hariharan, B.; Hayashi, Y.; Jagadeesan, P.; Jain, A.; Jhansi, V.B.; et al. Atmospheric temperature dependence of muon intensity measured by the GRAPES-3 experiment. In Proceedings of the 35th International Cosmic Ray Conference—ICRC2017, Busan, Korea, 12–20 July 2017.
20. Modzelewska, R.; Alania, M. Dependence of the 27-day variation of cosmic rays on the global magnetic field of the Sun. *Adv. Space Res.* **2012**, *50*, 716–724. [CrossRef]
21. Takai, H.; Feldman, C.; Minelli, M.; Sundermier, J.; Winters, G.; Russ, M.K.; Dodaro, J.; Varshney, A.; McIlwaine, C.J.; Tomaszewski, T.; et al. Tidal Frequencies in the Time Series Measurements of Atmospheric Muon Flux from Cosmic Rays. *arXiv* **2016**, arXiv:1610.05983.
22. Kovalyov, M.; Kovalyov, S. On the relationship between cosmic rays, solar activity and powerful earthquakes. *arXiv* **2014**, arXiv:1403.5728.

Article

Simulation and Evaluation of the Radiation Dose Deposited in Human Tissues by Atmospheric Neutrons

Ernesto Ortiz [1,2,*], Blanca Mendoza [1,3], Carlos Gay [2], Victor Manuel Mendoza [2], Marni Pazos [2] and Rene Garduño [2]

1 Escuela Nacional de Ciencias de la Tierra, Universidad Nacional Autónoma de México, Coyoacán, Ciudad de México 04510, Mexico; blanca@igeofisica.unam.mx
2 Centro de Ciencias de la Atmósfera, Universidad Nacional Autónoma de México, Coyoacán, Ciudad de México 04510, Mexico; cgay@unam.mx (C.G.); victor@atmosfera.unam.mx (V.M.M.); marni@unam.mx (M.P.); rene@atmosfera.unam.mx (R.G.)
3 Instituto de Geofísica, Universidad Nacional Autónoma de México, Coyoacán, Ciudad de México 04510, Mexico
* Correspondence: eortiz@encit.unam.mx

Citation: Ortiz, E.; Mendoza, B.; Gay, C.; Mendoza, V.M.; Pazos, M.; Garduño, R. Simulation and Evaluation of the Radiation Dose Deposited in Human Tissues by Atmospheric Neutrons. *Appl. Sci.* **2021**, *11*, 8338. https://doi.org/10.3390/app11188338

Academic Editors: Roberta Sparvoli and Francesco Palma

Received: 22 June 2021
Accepted: 31 August 2021
Published: 9 September 2021

Publisher's Note: MDPI stays neutral with regard to jurisdictional claims in published maps and institutional affiliations.

Copyright: © 2021 by the authors. Licensee MDPI, Basel, Switzerland. This article is an open access article distributed under the terms and conditions of the Creative Commons Attribution (CC BY) license (https://creativecommons.org/licenses/by/4.0/).

Abstract: The evaluation of the radiation dose (RD) deposited by atmospheric neutrons in human tissues is of vital importance due to the potential damages that over exposure to this radiation may cause to human health. The goal of this work was to obtain the RD that atmospheric neutrons with energy from 1 to 1000 MeV deposit in tissues of the human body (blood, adipose, bone and brain) as a function of both altitude and latitude. With the help of the Geant4 software, we developed a numerical simulation that allowed us to reach our goal; atmospheric neutron fluxes were obtained from the Excel-Based Program for Calculating Atmospheric Cosmic-Ray Spectrum (EXPACS). We found that the RD deposited by atmospheric neutrons increases with the increase in altitude and latitude, e.g., for an altitude of high mountain (4 km), the RD is increased ~19 times; while, for an altitude of commercial flights (10 km), the RD is increased ~156 times; in both cases, regarding the RD at sea level. We also found that, in the range of energies from 1 to 100 MeV, the RD deposited in the bone tissue sample is considerably lower that the RD deposited in the blood, adipose and brain tissue samples. On the other hand, for the range of energies between 200 and 1000 MeV, the RD deposited in the bone tissue sample is considerably greater that the RD deposited in the blood, adipose and brain tissue samples.

Keywords: atmospheric neutrons; radiation dose; passengers and flight crew; cosmic rays

1. Introduction

The atmospheric air showers are initiated by high-energy primary cosmic rays that enter the atmosphere isotropically from outer space, producing a large number of secondaries particles in a serie of successive collisions with target nuclei of the atmospheric constituents [1]. The primary cosmic rays which consists predominantly of protons, alpha particles and heavier nuclei are influenced by the galactic, the interplanetary, the magnetospheric and the geomagnetic magnetic fields while approaching the Earth [2].

In addition, the incoming cosmic rays with energies below about 20 GeV are modulated by the interplanetary magnetic field which is embedded in the solar wind [3], the expanding magnetized plasma generated by the Sun, which decelerates and partially excludes the lower energy cosmic rays from the inner solar system. There is a significant anticorrelation between solar activity and the intensity of the cosmic rays with the energies already mentioned.

Atmospheric neutrons are secondary particles resulting from interactions of primaries with nuclei of air constituents. It is useful to distinguish low-energy neutrons with energies less than about 10 MeV from high-energy neutrons. Low-energy neutrons are mostly evaporation products of excited nuclei and manifest an isotropic angular distribution. High-energy neutrons can be produced by hadronic collisions but also in charge exchange reactions of leading particles at very high energies, the angular distribution of high energy neutrons is anisotropic [2]. In addition, when high-energy particles collide with the atoms in the aircraft material, producing a local shower of particles, including electrons, lighter ions, gamma rays, and neutrons that are dangerous to human health [4].

Because the neutrons are produced by charged particles they inherit and also exhibit both latitude and longitude effects. Furthermore the flux of the low-energy neutron component is modulated by the solar cycle; in addition, strong solar flares may cause significant neutron fluxes in the atmosphere [2], sporadic enhancements of solar cosmic ray fluxes are also caused by coronal mass ejection, in these events the particles are accelerated to energies of hundreds of MeV up to several GeV [5,6].

The secondary energetic particles interacting with the human body may potentially cause an increase in cancer risk as the dose equivalent exposure increases, particularly in passengers and flight crew [7]. High RD may cause the breaks, potentially lethal to the cell by damaging the DNA strands, while low RD of ionizing radiations seems to have carcinogenic effects, even after years or decades, both in the exposed individuals and in subsequent generations [8]. The atmospheric neutron component of this complex radiation field, in particular, holds special interest in the cancer research community. At aircraft altitude (flight level around 10 km) of the total radiation dose deposited by secondary particles, the contribution of neutrons is around 40% [9].

The potential negative effect of atmospheric neutrons on human health is the motivation to carry out the present work, in this, we developed a numerical simulation using Geant4 software, with which, we calculated the radiation dose (RD) deposited by atmospheric neutrons in various tissues of the human body as function of the fluxes of these particles for various altitudes and latitudes; however, in all these geographical points we have maintained the same longitude (0°), since these will allow us to show the great variations of the RD as a function of altitude and latitude.

In this work, we have focused on the RD that atmospheric neutrons deposit in human tissues, if the readers want to know the effects of charged particles, they can read for example [6,8,10].

2. Materials and Methods: Setup of the Simulation

Excel-Based Program for Calculating Atmospheric Cosmic-Ray Spectrum (EXPACS) calculate cosmic ray fluxes of neutrons, protons, muons, electrons, positrons, and photons nearly anytime and anywhere in the Earth's atmosphere [11].

In Figure 1, we show the atmospheric neutron spectra for various altitudes obtained with EXPACS for geographical points with a latitude of 0° and a longitude of 0°. These data were calculated for the year 2020, period of minimum solar activity.

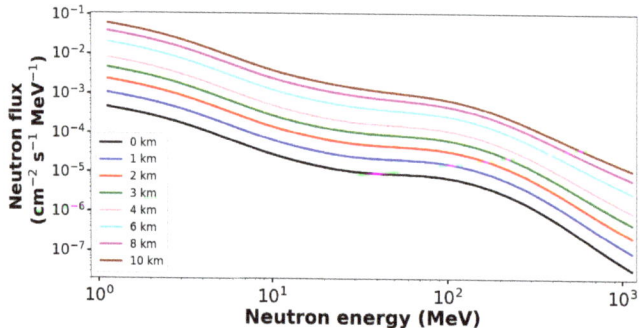

Figure 1. Atmospheric neutron spectra for various altitudes at geographical points with a latitude of 0° and a longitude of 0°. See text for more details.

In Figure 2, we show the atmospheric neutron spectra for various latitudes obtained with EXPACS for geographical points with an altitude of 10 km and a longitude of 0°. These data were calculated for the year 2020, period of minimum solar activity.

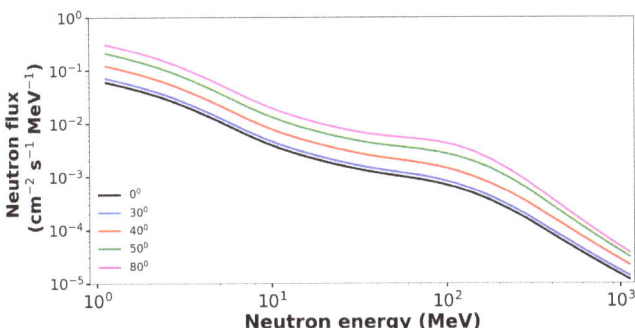

Figure 2. Atmospheric neutron spectra for various latitudes at geographical points with an altitude of 10 km and a longitude of 0°. See text for more details.

The Geant4 software is a toolkit for the simulation of the passage of particles through matter. It is used for a variety of applications domains, including high-energy physics, astrophysics and space science, medical physics and radiation protection [12]. Geant4 is used extensively in medical physics applications such as particles beam therapy, microdosimetry and radioprotection. The basic extensibility of the toolkit has facilitated its expansion into new user domains, such as biochemistry, material science and non-destructive scanning.

For this simulation we used as source, fluxes of 5×10^5 neutrons with kinetic energies between 1 and 1000 MeV. We also used tissues of blood (G4_BLOOD_ICRP, $\rho = 1.06$ g/cm^3), adipose (G4_ADIPOSE_TISSUE_ICRP, $\rho = 0.95$ g/cm^3), bone (G4_BONE_CORTICAL_ICRP, $\rho = 1.92$ g/cm^3) and brain (G4_BRAIN_ICRP, $\rho = 1.04$ g/cm^3) as samples obtained from Geant4 material's database, with a spherical geometry with a radius of 5 cm. These samples are tissues equivalent reported by The International Commission on Radiological Protection (ICRP). The ICRP is the primary body in protection against ionising radiation, it is a registered charity and is thus an independent non-governmental organisation created by the 1928 International Congress of Radiology to advance for the public benefit the science of radiological protection [13].

With the help of Geant4 software, we got the RD that neutrons deposit in the different sample tissues as a function of neutron's energy, the neutrons were injected isotropically to each of the samples. We arbitrarily choose geographic points, all of them with the same longitude (0°), at these points, we calculated the RD that neutrons deposited in the samples for different altitudes and different latitudes.

To obtain the RD in the tissues, we used the spectra shown in Figures 1 and 2 to calculate the neutron flux for each energy of interest, in addition, these fluxes were calculated for periods of one hour over the spherical area of the samples, later we multiplied the neutron flux and the RD deposited per neutron, for each energy, finally we added all products to obtain the RD deposited in the tissues.

3. Results and Discussion

When calculating the RD that neutrons deposit in the different sample tissues, under the conditions mentioned in the previous section, we obtained the statistical error less than 1% for each calculated value. Since statistical errors are relatively small, we decided not to point them out in Figure 3 that we present in this section.

In Figure 3, we show the RD that neutrons deposit in the samples as a funtion of their kinetic energy. In this figure, we can see three sections with different behavior of the RD deposited in the samples as a function of the energy of the neutrons; first, for energies of 1 to 20 MeV the RD curves show an increase with the energy of the neutrons; second, for energies of 20 to 500 MeV the RD curves are approximately flat; finally, for greater energies at 500 MeV the RD curves show an increase with the energy of the neutrons.

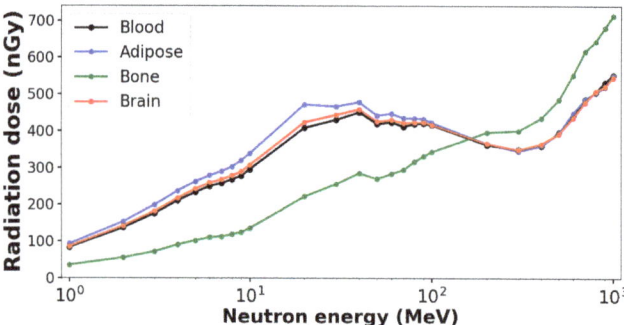

Figure 3. Radiation dose deposited by 5×10^5 neutrons to each energy in the tissue samples. This result was obtained by a numerical simulation.

In Figure 3, we can also observe that, neutrons with energy less than 200 MeV deposited a less RD in the bone sample compared with the RD deposited in the other samples, whereas that, the RD deposited in the adipose sample is greater than the RD deposited in blood and brain samples; this is due to the chemical composition of the tissues, for adipose tissue the dominant element is carbon, for blood and brain tissues is oxygen and for bone tissue are oxygen together with calcium. Neutrons with energy equal or greater than 200 MeV deposited a higher RD in the bone sample compared with the RD deposited in the other samples, this occurs because the cross section of the neutron varies with the kinetic energy of this particle.

Considering the RD deposited by the neutrons in the bone tissue sample as a reference, in the range of energies from 1 to 100 MeV, the RD deposited in the blood, adipose and brain tissue samples are 167.9%, 182.8% and 172.3% respectively; while, for the range of energies between 200 and 1000 MeV, the RD deposited in the blood, adipose and brain tissue samples are 81.0%, 80.8% and 80.5% respectively.

The neutron fluxes as a function of the altitude and the geographical latitude were calculated under the conditions mentioned in the previous section. In Figure 4, we show the RD that neutron with energies between 1 and 1000 MeV deposited in the tissue samples for an altitude between 0 and 10 km above sea level. To avoid overlapping of the curves, we have multiplied them by a factor shown in their respective labels.

In Figure 4, we can see that the RD deposited in the tissue samples is minimal at sea level and increases with height, the reason for this is that the neutron flux increases with increasing height. All curves have the same behavior. From the data obtained from the simulation and considering as reference the RD deposited by neutrons at 0 km above sea level, for an altitude of 4 km, corresponding approximately to an altitude of high mountain, the RD is increased ~19 times, while, for an altitude of 10 km, corresponding to an altitude of commercial flights, the RD is increased ~156 times.

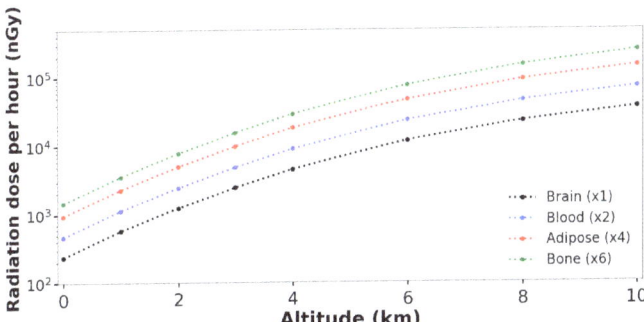

Figure 4. Radiation dose deposited in blood, adipose, brain and bone tissue samples by atmospheric neutrons with energies between 1 and 1000 MeV, as a function of the altitude for a latitude of 0° and a longitude of 0°. This result was obtained by a numerical simulation, the RD was calculated for a period of one hour in spherical samples with a radius of 5 cm. See text for more detail.

In Figure 5, we show the RD that neutron with energies between 1 and 1000 MeV deposited in the tissue samples as a function of the latitude for an altitude of 10 km above sea level and a longitude of 0°. To avoid overlapping of the curves, we have multiplied them by a factor shown in their respective labels. We can see that the RD deposited in the tissue samples is lower in equatorial zones (~0°); the above is due to the fact that in these zones the rigidity cutoff is maximum, containing lower flux of cosmic rays and consequences a lower production of atmospheric neutrons. In polar zones (~80°), the flux of cosmic rays is greater due to the low-rigidity cutoff which implies a greater production of atmospheric neutrons and consequently a higher RD deposited in the tissue samples. The RD in polar zones is approximately 4.3 times higher than the RD in equatorial zones for an altitud of 10 km.

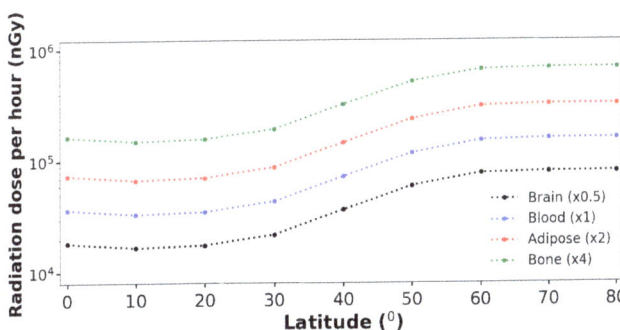

Figure 5. Radiation dose deposited in blood, adipose, brain and bone tissue samples by atmospheric neutrons with energies between 1 and 1000 MeV, as a function of the latitude for a altitude of 10 km and a longitude of 0°. This result was obtained by a numerical simulation, the RD was calculated for a period of one hour in spherical samples with a radius of 5 cm. See text for more detail.

The statistical errors of the data presented in Figures 4 and 5 are dominated by the propagation of the statistical error that is generated when calculating the neutron flux from the energy spectra, therefore, they present variations with the altitude and the latitude. We also found that statistical errors calculated as a percentage are the same for all tissue samples used.

The statistical errors of the RD deposited in each tissue sample of Figure 4 are 5.0%, 4.9%, 4.9%, 4.8%, 4.9%, 5.1%, 5.4% and 5.7% for altitudes of 0, 1, 2, 3, 4, 6, 8 and 10 km above sea level, respectively, while for the data in Figure 5 are 5.7%, 5.7%, 5.7%, 5.6%, 5.4%, 5.2%, 5.0%, 5.0% and 5.0% for latitudes of 0°, 10°, 20°, 30°, 40°, 50°, 60°, 70° and 80°, respectively.

4. Conclusions

In this work, and with the help of Geant4 software, first, we calculated the RD that neutrons with energy between 1 and 1000 MeV deposit in blood, adipose, bone and brain tissue samples; second, we calculated the RD that these particles deposit in the tissue samples as a function of different altitudes and latitudes, in both cases, for a longitude of 0°.

When we considered neutrons with a flat spectra (5×10^5 neutrons for each energy) and as a reference the RD deposited by them in the bone tissue sample, we found that in the range of energies from 1 to 100 MeV, the RD deposited in the blood, adipose and brain tissue samples is greater by 67.9%, 82.8% and 72.3% respectively. On the other hand, for the range of energies between 200 and 1000 MeV, the RD deposited in the blood, adipose and brain tissue samples is lower by 19.0%, 19.2% and 19.5% respectively.

When we obtained the atmospheric neutron spectra from [11]; we calculated the RD deposited by these particles in different human tissues as a function of both altitude and latitude and for a longitude of 0°. The results are shown in Figures 4 and 5. In general, we found that the RD deposited by atmospheric neutrons increases with the increase in altitude and latitude, e.g., for an altitude of 4 km (altitude of high mountain), the RD is increased ~19 times; while, for an altitude of 10 km (altitude of commercial flights), the RD is increased ~156 times, respectively; in both cases, regarding the RD at sea level.

Author Contributions: Conceptualization, E.O. and B.M.; Investigation, E.O.; Methodology, E.O.; Software, E.O.; Supervision, B.M. and C.G.; Writing—original draft, E.O.; Writing—review & editing, B.M., C.G., V.M.M., M.P. and R.G. All authors have read and agreed to the published version of the manuscript.

Funding: This research was funded by Posdoctoral Grant Programa de Becas Posdoctorales de la Universidad Nacional Autónoma de México.

Institutional Review Board Statement: Not applicable, for studies not involving humans or animals.

Informed Consent Statement: Not applicable, for studies not involving humans or personal data.

Data Availability Statement: Data are available, contact the corresponding author.

Acknowledgments: The authors thank The European Organization for Nuclear Research (CERN) for Geant4 software.

Conflicts of Interest: The authors declare no conflict of interest. The sponsors had no role in the design, execution, interpretation, or writing of the study.

References

1. Grieder, P.K.F. *Extensive Air Showers. High Energy Phenomena and Astrophysical Aspects. A Tutorial, Reference Manual and Data Book*; Springer: Berlin/Heidelberg, Germany, 2010; pp. 6–16.
2. Grieder, P.K.F. *Cosmic Rays at Earth: Researcher's Reference Manual Data Book*; Elsevier Science B.V.: Amsterdam, The Netherlands, 2001; pp. 77–100.
3. Valdés-Galicia, J.F.; González, L.X. Solar modulation of low energy galactic cosmic rays in the near-earth space environment. *Adv. Space Res.* **2016**, *57*, 1294–1306. [CrossRef]
4. Warden, D.; Bayazitoglu, Y. Consideration of backscatter radiation from galactic cosmic rays in spacecraft shielding design. *J. Thermophys. Heat Transf.* **2021**, *35*. [CrossRef]
5. Gil, A.; Kovaltsov, G.A.; Mikhailov, V.V.; Mishev, A.; Poluianov, S.; Usoskin, I.G. An anisotropic cosmic-ray enhancement event on 07-June-2015: A possible origin. *So. Phys.* **2018**, *293*, 154. [CrossRef]
6. Belisheva, N.K.; Lammer, H.; Biernat, H.K.; Vashenuyk, E.V. The effect of cosmic rays on biological systems—An investigation during GLE events. *Astrophys. Space Sci. Trans.* **2012**, *8*, 7–17. [CrossRef]
7. Tobiska, W.K.; Atwell, W.; Beck, P.; Benton, E.; Copeland, K.; Dyer, C.; Gersey, B.; Getley, I.; Hands, A.; Holland, M.; et al. Atmospheric radiation measurements and modeling needed to improve air safety. *Space Weather* **2015**, *13*, 202–210. [CrossRef]
8. Burgio, E.; Piscitelli, P.; Migliore, L. Ionizing radiation and human health: Reviewing models of exposure and mechanisms of cellular damage. An epigenetic perspective. *Int. J. Environ. Res. Public Health* **2018**, *15*, 1971. [CrossRef] [PubMed]

9. Vulković, B.; Poje, M.; Varga, M.; Radolić, V.; Miklavčić, I.; Faj, D.; Stanić, D.; Planinić, J. Measurements of neutron radiation in aircraft. *Appl. Radiat. Isot.* **2010**, *68*, 2398–2402. [CrossRef] [PubMed]
10. Cekanaviciute, E.; Rosi, S.; Costes, S.V. Central nervous system responses to simulated galactic cosmic rays. *Int. J. Mol. Sci.* **2018**, *19*, 3669. [CrossRef] [PubMed]
11. EXcel-Based Program for Calculating Atmospheric Cosmic-Ray Spectrum (EXPACS). Available online: https://phits.jaea.go.jp/expacs/ (accessed on 21 June 2021).
12. Allison, J.; Amako, K.; Apostolakis, J.; Arce, P.; Asai, M.; Aso, T.; Bagli, E.; Bagulya, A.; Banerjee, S.; Barrand, G.; et al. Recent developments in Geant4. *Nucl. Instrum. Methods Phys. Res. A* **2016**, *835*, 186–225. [CrossRef]
13. International Commission on Radiological Protection. *ICRP Publication 119. Compendium of Dose Coefficients Based on ICRP Publication 60*; Ann. ICRP 41; Elsevier: Amsterdam, The Netherlands, 2012.

Article

Direct Measurements of Cosmic Rays (TeV and beyond) Using an Ultrathin Calorimeter: Lessening Fluctuation Method

Igor Lebedev [1,*], Anastasia Fedosimova [2,*], Andrey Mayorov [3], Pavel Krassovitskiy [4], Elena Dmitriyeva [1], Sayora Ibraimova [1] and Ekaterina Bondar [1]

[1] Institute of Physics and Technology, Satbayev University, Almaty 050032, Kazakhstan; dmitriyeva2017@mail.ru (E.D.); sayara_ibraimova@mail.ru (S.I.); grushevskaiya@bk.ru (E.B.)
[2] Faculty of Physics and Technology, Al-Farabi Kazakh National University, Almaty 050040, Kazakhstan
[3] Department of Experimental Nuclear and Space Physics, Moscow Engineering and Physics Institute, National Research Nuclear University MEPhI, 115409 Moscow, Russia; agmayorov@mephi.ru
[4] Laboratory of Theoretical Nuclear Physics, Institute of Nuclear Physics, Almaty 050032, Kazakhstan; pavel.kras@inp.kz
* Correspondence: lebedev692007@yandex.ru (I.L.); ananastasia@list.ru (A.F.)

Abstract: In this paper, we propose a method that makes it possible to use an ultrathin calorimeter for direct measurements of cosmic rays with energies of TeV and higher. The problems of determining the primary energy with a thin calorimeter, due to large fluctuations in shower development, the low statistics of analyzed events and the large size required for the calorimeter, are considered in detail. A solution to these problems is proposed on the basis of a lessening fluctuation method. This method is based on the assumption of the universality of the development of cascades initiated by particles of the same energy and mass. For energy reconstruction, so-called correlation curves are used. The main analyzed quantities are the size of the cascade and the rate of its development. The method was tested using the calorimeter of the PAMELA collaboration. Based on simulations, it is shown that the primary energy can be determined on the ascending branch of the cascade curve. This fact solves the problems associated with the need to increase the calorimeter thickness with an increase in primary energy and with the limitation of the analyzed events. The proposed technique is universal for different energies and different nuclei.

Keywords: ultrathin calorimeter; cosmic rays; direct measurements; energy reconstruction; PAMELA; shower development universality

1. Introduction

Measurements of the chemical composition and fluxes of cosmic rays play a decisive role in understanding the mechanisms of their acceleration and propagation. Different cosmological models predict different elemental composition of cosmic rays and different spectra of the elements [1–3].

Cosmic rays at energies E > 100 TeV are studied at ground-based cosmic ray stations based on the analysis of extensive air showers [4,5]. Cosmic rays at E < 100 TeV are studied by direct measurements outside the Earth's atmosphere on spacecraft or high-altitude aerostats. The main advantage of direct experiments is the ability to measure the charge of the incident particle.

The energies of cosmic particles are measured fairly accurately for particles with energies E < 100 GeV. Modern magnetic spectrometers can detect the primary energy with an error of less than 10 percent. Such devices have limitations at energies of TeV and higher [6–8].

In the region of 1–100 TeV, there is a lack of experimental methods. Today, there is practically only one reliable method for measuring the energy of various nuclei at energies of TeV and above: this is the ionization calorimeter method [9,10].

At present, calorimeters are used in almost all experiments in the study of cosmic rays, in which the equipment is placed on high-altitude balloons or spacecraft. The main problem with this method of measuring energy is that it requires heavy devices, since the calorimeter must have sufficient depth to determine the value of the total energy release in the calorimeter. Moreover, the higher the primary energy, the thicker the calorimeter should be. The huge weight of the installation makes it much more difficult to use such a device in space experiments.

A more promising approach to determining the energy of cosmic rays based on direct measurements is the use of a thin calorimeter. In a thin calorimeter, the entire cascade of secondary particles is not recorded, but only the beginning of the cascade is measured. Many methods have been developed for measuring the energy of the initial particles using various types of thin calorimeters. However, due to significant fluctuations in the development of the cascade, the energy resolution of thin calorimeters when measuring hadron cascades at the present stage is 30–70% [11–19].

Most of the energy measurement methods used in modern experiments are based on the use of a cascade curve—the dependence of the cascade size (usually, the logarithm of the energy release, $\log q$, at the calorimeter measurement layer is used) on the penetration depth (d) of cascade to this measurement layer.

If the cascade curve has reached its maximum in the calorimeter, then the primary energy is reconstructed quite accurately. However, in order to measure the maximum of the cascade, the calorimeter must have a sufficiently large thickness. Moreover, the higher the primary energy, the thicker the calorimeter should be.

If the maximum of the cascade curve is not reached in the calorimeter, then the energy release at the last layer of the calorimeter, or the total energy release in the calorimeter, is used to determine the energy. Cascade curves fluctuate significantly. The cascade can begin to develop on the first measurement layer, for example, or on the 10th measurement layer. Accordingly, the total energy release in the calorimeter for these two cascades will differ significantly. Furthermore, since total energy release is used to define primary energy, large fluctuations in total energy release lead to large errors in primary energy reconstruction.

This paper presents a lessening fluctuation method (LFM) to improve energy reconstruction for data obtained with thin calorimeters. The proposed method is based on the use of so-called correlation curves—the dependence of the cascade size ($S = \log q$) on the cascade development rate (R). The cascade development rate is understood as a value equal to the difference in the cascade size at two measurement levels, divided by the calorimeter thickness, during the passage of which this change in the cascade size occurs: $R = (S_1 - S_2)/(d_1 - d_2)$, where d_1 and d_2 are the penetration depths to these two measurement layers. The cascade development rate depends on the primary energy and, therefore, can be used as an additional value to improve the accuracy of the reconstruction of the primary energy. The size–rate curves practically do not fluctuate. They coincide with the cascade, which begins to develop at the first measuring layer, and the cascade, which begins at the 10th measuring layer. Therefore, the energy resolution is better than using cascade curves.

Moreover, using this method, the primary energy is reconstructed near the beginning of the development of the cascade. Thus, with increasing energy, it is not necessary to increase the calorimeter thickness. Moreover, the calorimeter thickness can be reduced and an ultrathin calorimeter used.

2. Primary Energy Measurement with a Calorimeter

The technical realization of modern ionization calorimeters can be variable, but the idea remains invariable: the primary particle enters into a dense substance (absorber), in which numerous nuclear and electromagnetic interactions take place. It gives rise to a cascade of secondary particles. To measure the characteristics of the cascade, the dense substance is sandwiched with special detectors. By measurement of signals from these detectors, the cascade curve is formed.

By their design, calorimeters are divided into homogeneous and heterogeneous. Heterogeneous calorimeters consist of layers of a substance with a high density (lead, tungsten), where particles lose their energy during passage, alternating with layers of detectors (silicon), where the energy released by the particles of the cascade is measured. Homogeneous calorimeters use substances (bismuth germanate crystal, lead tungstate, etc.) which are simultaneously both an absorber and a detector.

The geometric dimensions of heterogeneous calorimeters are usually significantly lower than those of homogeneous ones. In addition, they have better spatial resolution as they are segmented in both the longitudinal and lateral directions. A significant drawback of heterogeneous calorimeters is the transient effect due to the significant difference in the densities of the absorber and detector. Cascade curves develop differently in different materials. Consequently, the behavior of the cascade curve is violated when the cascade transitions from one material to another. In this regard, fluctuations in the development of the cascade from layer to layer can be observed. The strongest fluctuations from layer to layer are at the beginning of the development of the cascade. This makes the analysis very difficult [20].

Several methods have been developed to measure the primary energy of cosmic rays using different types of calorimeters.

The PAMELA calorimeter is a heterogeneous calorimeter. It consists of 22(x, y) silicon detector planes alternating with tungsten absorber planes. The calorimeter thickness is 16.3 radiation lengths. In the PAMELA experiment, the primary energy is estimated from the maximum of the cascade curve describing the longitudinal profile of the shower developed in the calorimeter. If the shower maximum is located outside the calorimeter, then the energy released in the last layer of the calorimeter is used to estimate the energy. This technique provides an energy resolution for protons of ~40% [14].

The NUCLEON calorimeter thickness is 15.3 radiation lengths (the silicon microstrip detectors interleaved with thin tungsten layers). The proposed technique for primary CR energy measurement is based on the generalized Castagnoli kinematical method (KLEM method) developed for emulsion. In this method, the primary energy is reconstructed by registering the spatial density of the secondary particles. Secondary particles are generated by the first hadronic inelastic interaction in a carbon target. Additional particles are then produced in a thin tungsten converter by electro-magnetic and hadronic interactions. This method provides an energy resolution of ~70% [16].

CALET is a homogeneous calorimeter made of lead tungstate ($PbWO_4$) bars arranged in 12 layers. The total thickness of the device is equivalent to 30 radiation lengths. The primary particle energy is calculated from the total energy release in the calorimeter. The energy released in the calorimeter is scaled linearly with the energy of the incident particle. The obtained energy resolution is close to 30% [17].

ATIC is a homogeneous calorimeter consisting of 10 layers of 40 bismuth germanium scintillation crystals (BGO). The total thickness is approximately 22 radiation lengths. For protons, most of the released energy is not recorded by the calorimeter detectors. In this regard, the selection of events was carried out according to predetermined conditions, such as the interaction near the upper boundary of the calorimeter. Despite this, the energy resolution for protons is ~30% due to large fluctuations in the energy release of the hadron cascade [11].

DAMPE is a homogeneous calorimeter of about 31.5 radiation lengths. The calorimeter is made on the basis of bismuth germanate crystals. For energy reconstruction, MC simulations are used to derive the energy response matrix, applying some selections. Then a deconvolution of the measured energy distribution into the incident energy distribution is applied. The number of events in the i-th deposited energy bin is obtained via the sum of the number of events in all the incident energy bins weighted by the energy response matrix. The energy resolution for protons is approximately 35% [18].

Figure 1 shows the proton energy spectra measured in various experiments. As can be seen from Figure 1, at energies up to 100 GeV, all the presented spectra practically coincide.

At energies higher than 100 GeV, the difference becomes more significant, and the spectrum measurement errors are significantly higher.

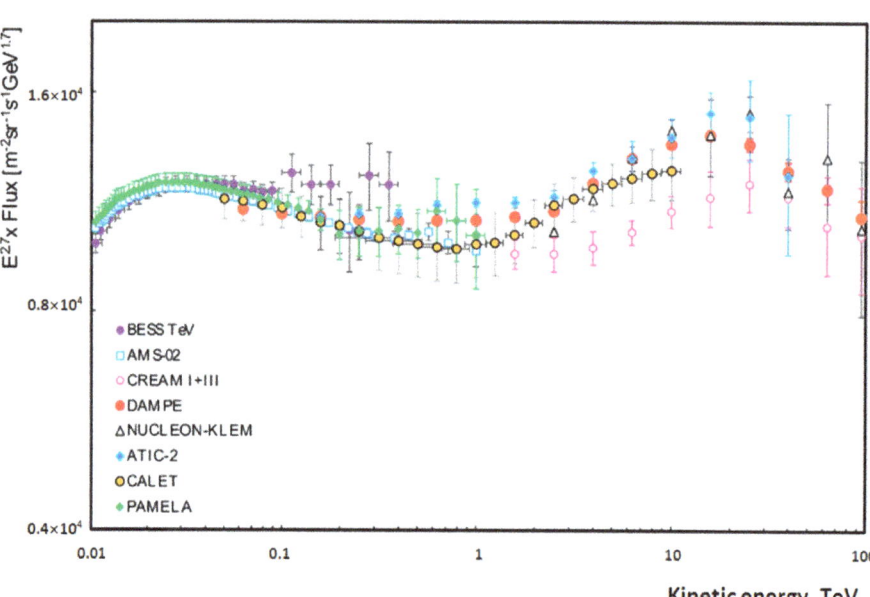

Figure 1. Proton spectra of various experiments [11–19].

Modern experiments with calorimeters for 20–30 radiation lengths do not allow measuring the entire flux of cosmic rays. Many events do not reach the maximum of the cascade and they are excluded from the analysis. Alternatively, their energy is determined with low accuracy. In this regard, the real energy spectrum is distorted. The higher the energy of the primary particle, the thicker the calorimeter should be. To lift a much larger calorimeter into space requires enormous financial cost. Therefore, it is necessary to look for ways to reduce the influence of fluctuations in the development of the cascade on the results of measuring the primary energy.

3. Lessening Fluctuation Method

LFM has been tested on the PAMELA calorimeter. Simulation of the development of cascade processes formed by primary particles of various masses and energies was carried out using the GEANT4 10.4 software package [21].

3.1. Fluctuations in Cascade Development

To determine the primary energy E based on the energy release at the observation level, usually the following dependence is used: $q = aE^b$, where a, b are parameters depending on the penetration depth d and the mass of the primary particle. The equation is statistically correct. However, q (at the observation level) strongly fluctuates in an individual event. Therefore, in order to solve the problem of large fluctuations in the development of a cascade, it is necessary to start with an analysis of fluctuations in individual events. For this, we first considered individual cascades with significantly different cascade curves.

As the analyzed value of the cascade size, we used the logarithm of the energy release, $\log q$, at the measuring layer. The measuring levels of the PAMELA calorimeter were equidistant. Therefore, as the cascade curved, we analyzed the dependence of the cascade size on the layer number, L.

The cascade, the development of which starts at the beginning of the calorimeter, is called a fast cascade. The cascade, the significant development of which begins in the second half of the calorimeter, is called a slow cascade.

Figure 2a shows the cascade curves for three cascades of 10 TeV protons with fast, medium, and slow character of the development of the cascade process.

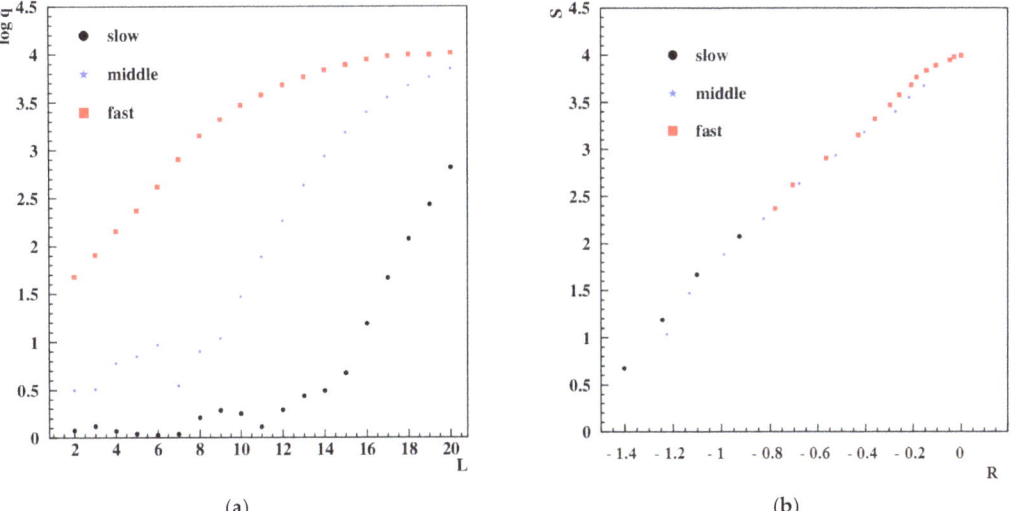

Figure 2. (a) Cascade curves for three showers of 10 TeV protons with fast, medium, and slow character of the development of the cascade process; (b) The size–rate curves for the same showers.

As can be seen in Figure 2a, the total energy release in the calorimeter for these three showers differs significantly. The total energy release by the fast shower is several times higher than that released by the slow shower. Therefore, methods using this parameter to reconstruct the primary energy will determine that the energy of these showers is significantly different, while the primary energy of these showers is the same. The situation is similar for methods that use the energy release on the last calorimeter layer as a calculated value, since this value also differs significantly for these three showers.

Determination of the energy from the maximum of the cascade curve is possible only for fast showers. Slow and medium showers cannot be analyzed because they do not reach the maximum of the cascade curve. Thus, the statistics of the analyzed events are significantly reduced. Moreover, the maximum point shifts towards greater depth with increasing energy. Therefore, the higher the primary energy, the thicker the calorimeter should be.

To understand how to solve the problem of large fluctuations in the development of a cascade, it is necessary to understand what their causes are.

The behavior of the cascade curve depends on the features of the interaction of the primary nucleus with the nuclei of the calorimetric substance. Different approaches and methods are used in order to study features in the multi-particle production [22–26]. First, the fluctuations of the penetration depth before the first interaction are very important. The earlier the primary particle interacts, the faster the cascade begins to develop. Second, fluctuations in the multiplicity of the first interaction are important. If the first interaction is central, then many particles are produced and the cascade develops rapidly. If the first interaction is peripheral, then the cascade develops slowly. For example, the proton that initiated the middle cascade (Figure 2a) interacted at the beginning of the calorimeter, but the rapid development of the cascade began only after layer 9.

After the first interaction, each secondary particle can interact, also producing secondary particles. The parameters (the penetration depth before the first interaction, the number of secondary particles, etc.) of each subsequent interaction also fluctuate. However, since there are several particles in a cascade, fluctuations of individual interactions can partially compensate each other. When there are many particles in a cascade, the property of universality of the cascade development is realized. All cascades initiated by particles of the same energy and mass develop in the same way [27,28].

Thus, the cascades differ greatly in the depth of the first interaction. They also differ greatly in the parameters of the first interaction (multiplicity, peripherality, etc.). This leads to major fluctuations at the beginning of the cascade curve. However, the cascade fluctuates weakly if it contains many secondary particles. Thus, instead of the penetration depth, it is necessary to find another parameter that does not depend on these fluctuations in the development of the cascade process.

LFM is based on the use of so-called correlation curves—the dependence of the cascade size ($S = \log q$) on the cascade development rate $R = (S_1 - S_2)/(d_1 - d_2)$. As unit absorber ($d_1 - d_2$), a thickness equal to three layers of the PAMELA calorimeter was chosen. Therefore, the rate of shower development was calculated as the difference between the cascade size on the L-th and $L + 3$ measuring layers, $R = S_L - S_{L+3}$.

When choosing the thickness of the unit absorber, we took into account two main factors. The first factor was fluctuations from layer to layer. The thinner the unit absorber, the higher the relative fluctuations of R due to fluctuations from layer to layer. Therefore, it was preferable to choose a thicker unit absorber. The second factor was the calorimeter thickness. The thinner the unit absorber used in the LFM, the more points on the correlation curve could be obtained. Looking to the future, a thin unit absorber makes it possible to use an ultra-thin calorimeter to measure primary energy.

Figure 2b shows the correlation curves for the same three showers that were presented in Figure 2a. As can be seen in Figure 2b, in contrast to the cascade curves (Figure 2a), the correlation curves almost coincide for fast, medium, and slow showers.

This greatly simplifies the task of determining the primary energy. Regardless of fluctuations in the development of a shower, all proton showers of the same energy are located on the same curve. Therefore, the energy of these showers will also be defined as the same.

3.2. The Analysis Procedure: The Size–Rate Function

The analysis procedure consisted of several main stages.

First stage: simulation of cascades with fixed energies.

We simulated 100 cascades initiated by iron nuclei, 100 carbon cascades and 100 proton cascades with fixed energies of 1 TeV and 10 TeV in the PAMELA calorimeter.

The mean cascade curves for these cascades are presented in Figure 3a. Error bars show statistical errors. As can be seen in Figure 3a, the development of the proton and Fe cascades is significantly different. At $L < 8$ Fe showers with energy of 1 TeV have $\log q$ higher than proton showers with energy of 10 TeV. The most significant factors determining the observed differences are the penetration depth before the first interaction and the number of secondary particles formed.

Second stage: smoothing the cascade curve.

To reduce fluctuations from layer to layer, a signal accumulation method along the spectrum was used. This method allows for a softer minimization of fluctuations from layer to layer, in contrast, for example, to fitting with a polynomial function [29]. Smoothing was carried out at three points in accordance with the formula:

$$S_L = \frac{1}{3} \sum_{i=L-1}^{L+1} \log q_i$$

where log q_i is the measured value of the shower size, and S_L is the accumulated value of the shower size.

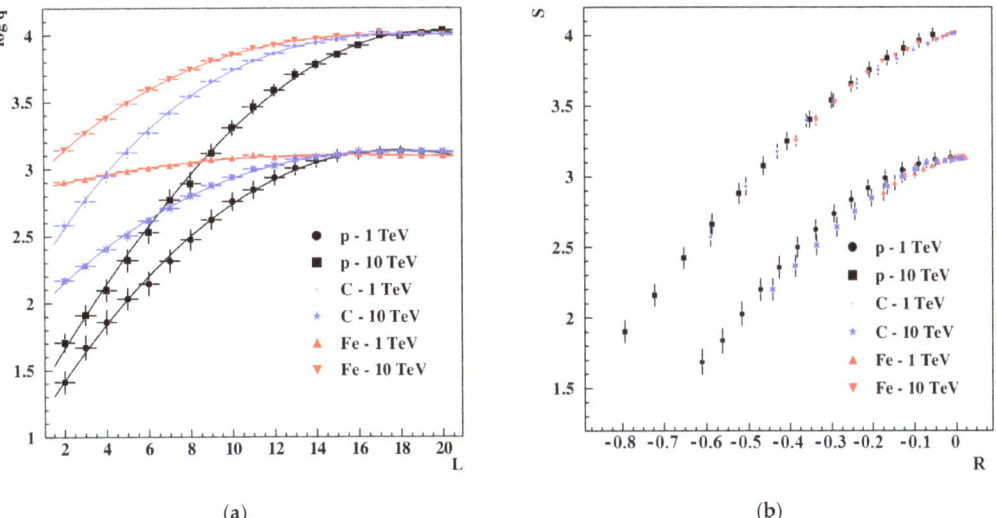

Figure 3. (a) Cascade curves for showers initiated by protons, C and Fe nuclei with energies of 1 TeV and 10 TeV in the PAMELA calorimeter; (b) The size–rate curves for the same showers.

Third stage: searching for the beginning of the cascade development.

As can be seen in Figure 3a, the avalanche-like process of the cascade development does not always begin immediately after the interaction of the primary particle with the calorimeter substance. If the first interaction is peripheral, then the energy release from layer to layer can not only increase, but also decrease. We considered such cascades as not having started. In order to separate the part of the measurements in which the cascade had not yet begun, we compared the rate of development of the cascade in adjacent layers. If the rate of development of a cascade increased at three adjacent observation levels in a row, then such a cascade was considered to have begun. For example, the fast cascade shown in Figure 3a was considered to have started at the first level, the middle at layer 9, and the slow cascade only at layer 15.

Fourth stage: plotting SR distributions.

Figure 3b shows the average size–rate dependences for the same proton, carbon and Fe cascades as in Figure 3a.

As can be seen in Figure 3b the size–rate dependences are an ordered structure depending on the primary energy and are practically independent of the type of the primary nucleus. This fact can also be attributed to the advantages of the presented approach.

Fifth stage: creating the size–rate function.

To create the SR function, we fitted the SR curves of third-order polynomial functions for each fixed energy:

$$S(R) = a_0 + a_1 R + a_2 R^2 + a_3 R^3 \qquad (1)$$

Then the coefficients a_0, a_1, a_2, a_3 were fitted depending on the energy. The size–rate function for reconstructing primary energy was in the following form:

$$S(R,E) = a_0(E) + a_1(E)R + a_2(E)R^2 + a_3(E)R^3 \qquad (2)$$

Using the size–rate function (2), an analysis of test cascades was performed.

3.3. Analysis of Test Cascades

For the analysis 100 test cascades formed by primary protons, 100 carbon cascades and 100 cascades formed by iron nuclei with random energies in the range from 1 TeV to 10 TeV were simulated.

Reconstruction of the primary energy was based on dependence (2). In order to determine the energy of the i-th test cascade using (2), it was necessary to substitute into the function (2) the "measured" value of the rate R_m and to vary E_{rec} in order to minimize the difference between "measured" value of the size S_m and the size–rate function (2):

$$| S_m - S(R_m, E_{rec}) | = | S_m - (a_0(E_{rec}) + a_1(E_{rec}) R_m + a_2(E_{rec}) R_m^2 + a_3(E_{rec}) R_m^3) | < \varepsilon$$

In the calculations, we used $\varepsilon = 0.001$. Energy reconstruction errors were calculated using the formula:

$$\sigma E = \sqrt{\frac{\sum (E - E_{rec})^2}{n-1}},$$

where n is number of reconstructed events.

Figure 4 shows the energy resolution reached by this procedure. The energy resolution is practically independent of the primary energy.

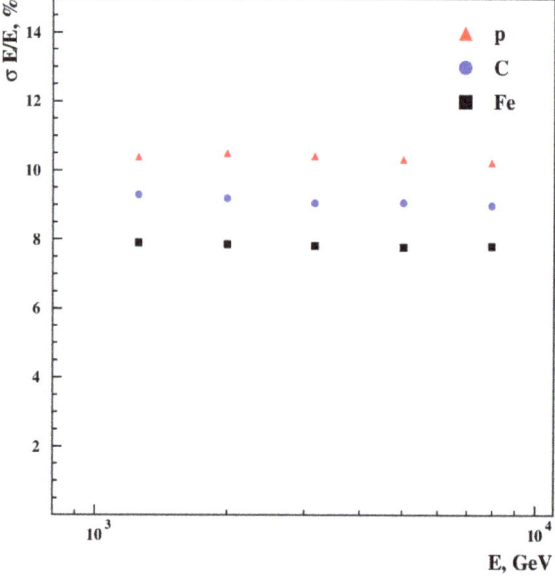

Figure 4. Energy resolution for p, C and Fe cascades at different energies in the PAMELA calorimeter.

In this paper, we did not consider the correlation between fit parameters and how they affect the ultimate energy reconstruction. It is likely that an improvement in the fitting parameters can improve the results of primary energy reconstruction.

The correlation curves presented in this paper were constructed for analysis of the PAMELA calorimeter. In case of a change in the geometry and material of the calorimeter, the analysis procedure must be repeated in full.

4. Conclusions

In this paper, we considered the possibility of using an ultrathin calorimeter for direct measurements of cosmic rays with energies TeV and higher. The following problems of measuring the energy of cosmic particles using a thin calorimeter were considered in detail: large fluctuations in the development of cascade processes (they lead to significant errors in determining the energy); limiting the number of analyzed events (particles whose cascade curves have not reached their maximum cannot be measured and they are excluded from the analysis); large calorimeter sizes (the higher the primary energy, the thicker the calorimeter should be). A solution to these problems is proposed on the basis of a lessening fluctuation method. This method is based on the assumption of the universality of the development of cascades formed by particles of the same energy and charge. For energy reconstruction, so called, SR curves are used. The main analyzed quantities are: S—the size of the cascade (the energy deposited on each layer of the calorimeter); R—the rate of development of the cascade (the difference in the cascade size on two measuring layers of the calorimeter).

Based on simulations of the PAMELA calorimeter, it is shown that the SR curves are almost parallel to each other and practically do not depend on the depth of the cascade development. It makes it possible to determine the primary energy for cascades that have not reached their maximum. This fact solves the problem associated with the need to increase the calorimeter thickness with increasing primary energy. Therefore, an ultrathin calorimeter can be used for measurement. In addition, the statistics of the analyzed events can be increased. Correlation curves fluctuate much less than cascade curves. Therefore, the energy resolution for protons is improved by ~10 percent. The proposed technique is universal for different energies and different nuclei.

Author Contributions: Conceptualization and methodology, I.L. and A.F.; software and simulation, A.M.; analysis and investigation, E.D., S.I., P.K. and E.B. All authors have read and agreed to the published version of the manuscript.

Funding: This research was funded by the Ministry of Education and Science of the Republic of Kazakhstan grant number AP08855403.

Conflicts of Interest: The authors declare no conflict of interest.

References

1. Ptuskin, V.; Zirakashvili, V.; Seo, E.S. Spectra of cosmic-ray protons and helium produced in supernova remnants. *Astrophys. J.* **2013**, *763*, 47. [CrossRef]
2. Boos, E.; Lebedev, I.; Philippova, L. Modelling of an energy spectrum of cosmic rays in the knee region. *J. Phys. G Nucl. Part Phys.* **2006**, *32*, 2273–2278. [CrossRef]
3. Erlykin, A.D.; Wolfendale, A.W. Interpretation of features in the cosmic ray proton and helium energy spectra in terms of a local source. *J. Phys. G Nucl. Part. Phys.* **2015**, *42*, 125201. [CrossRef]
4. Kuzmichev, L.; Astapov, I.; Bezyazeekov, P.; Borodin, A.; Brückner, M.; Budnev, N.; Chiavassa, A.; Gress, O.; Gress, T.; Grishin, O.; et al. Cherenkov EAS arrays in the Tunka astrophysical center: From Tunka-133 to the TAIGA gamma and cosmic ray hybrid detector. *Nucl. Instrum. Methods Phys. Res. Sect. A Accel. Spectrometers Detect. Assoc. Equip.* **2020**, *952*, 161830. [CrossRef]
5. Apel, W.D.; Badea, A.; Bekk, K.; Blümer, J.; Boos, E.; Bozdog, H.; Brancus, I.; Daumiller, K.; Doll, P.; Engel, R. Applying shower development universality to KASCADE data. *Astropart. Phys.* **2008**, *29*, 412–419. [CrossRef]
6. Aguilar, M.; Cavasonza, L.A.; Alpat, B.; Ambrosi, G.; Arruda, L.; Attig, N.; Aupetit, S.; Azzarello, P.; Bachlechner, A.; Barao, F.; et al. Precision measurement of cosmic-ray nitrogen and its primary and secondary components with the alpha magnetic spectrometer on the International Space Station. *Phys. Rev. Lett.* **2018**, *121*, 051103. [CrossRef]
7. Mocchiutti, E. Direct detection of cosmic rays: Through a new era of precision measurements of particle fluxes. *Nucl. Phys. B* **2014**, *256–257*, 161–172. [CrossRef]
8. Adriani, O.; Barbarino, G.; Bazilevskaya, G.; Bellotti, R.; Boezio, M.; Bogomolov, E.; Bongi, M.; Bonvicini, V.; Bottai, S.; Bruno, A.; et al. The PAMELA Mission: Heralding a new era in precision cosmic ray physics. *Phys. Rep.* **2014**, *544*, 323. [CrossRef]
9. Borisov, S.V.; Voronov, S.A.; Karelin, A.V. Energy measurements of electrons and protons in cosmic ray physics using satellite and balloon calorimeters in recent two decades. *Cosm. Res.* **2011**, *49*, 247. [CrossRef]
10. Sparvoli, R. Direct measurements of cosmic rays in space. *Nucl. Phys. B Proc. Suppl.* **2013**, *239–240*, 115–122. [CrossRef]
11. Panov, A.D.; Adams, J.H.; Ahn, H.S.; Batkov, K.E.; Bashindzhagyan, G.L.; Watts, J.W.; Wefel, J.P.; Wu, J.; Ganel, O.; Guzik, T.G.; et al. Elemental energy spectra of cosmic rays from the data of the ATIC-2 experiment. *Bull. Russ. Acad. Sci. Phys.* **2007**, *71*, 494. [CrossRef]
12. Yoon, Y.S.; Ahn, H.S.; Allison, P.S.; Bagliesi, M.G.; Beatty, J.; Bigongiari, G.; Boyle, P.J.; Childers, J.T.; Conklin, N.B.; Coutu, S.; et al. Cosmic-ray proton and helium spectra from the first CREAM flight. *Astrophys. J.* **2011**, *728*, 122. [CrossRef]

13. Yoon, Y.S.; Anderson, T.; Barrau, A.; Conklin, N.B.; Coutu, S.; Derome, L.; Han, J.H.; Jeon, J.A.; Kim, K.C.; Kim, M.H.; et al. Proton and Helium Spectra from the CREAM-III Flight. *Astrophys. J.* **2017**, *839*, 5. [CrossRef]
14. Adriani, O.; Barbarino, G.C.; Bazilevskaya, G.A.; Bellotti, R.; Boezio, M.; Bogomolov, E.A.; Bongi, M.; Bonvicini, V.; Bottai, S.; Bruno, A.; et al. Ten years of PAMELA in space. *Riv. Nuovo Cim.* **2017**, *40*, 1.
15. Aguilar, M.; Aisa, D.; Alpat, B.; Alvino, A.; Ambrosi, G.; Andeen, K.; Arruda, L.; Attig, N.; Azzarello, P.; Bachlechner, A.; et al. Precision Measurement of the Proton Flux in Primary Cosmic Rays from Rigidity 1 GV to 1.8 TV with the Alpha Magnetic Spectrometer on the International Space Station. *Phys. Rev. Lett.* **2015**, *114*, 171103.
16. Atkin, E.; Bulatov, V.; Dorokhov, V.; Gorbunov, N.; Filippov, S.; Grebenyuk, V.; Karmanov, D.; Kovalev, I.; Kudryashov, I.; Kurganov, A.; et al. New Universal Cosmic-Ray Knee near a Magnetic Rigidity of 10 TV with the NUCLEON Space Observatory. *JETP Lett.* **2018**, *108*, 5. [CrossRef]
17. Adriani, O.; Akaike, Y.; Asano, K.; Asaoka, Y.; Bagliesi, M.G.; Berti, E.; Bigongiari, G.; Binns, W.R.; Bonechi, S.; Bongi, M.; et al. Direct Measurement of the Cosmic-Ray Proton Spectrum from 50 GeV to 10 TeV with the Calorimetric Electron Telescope on the International Space Station. *Phys. Rev. Lett.* **2019**, *122*, 181102. [CrossRef]
18. An, Q.; Asfandiyarov, R.; Azzarello, P.; Bernardini, P.; Bi, X.J.; Cai, M.S.; Chang, J.; Chen, D.Y.; Chen, H.F.; Chen, J.L.; et al. Measurement of the cosmic ray proton spectrum from 40 GeV to 100 TeV with the DAMPE satellite. *Sci. Adv.* **2019**, *5*, aax3793.
19. Haino, S.; Sanuki, T.; Abe, K.; Anraku, K.; Asaoka, Y.; Fuke, H.; Imori, M.; Itasaki, A.; Maeno, T.; Makida, Y.; et al. Measurements of primary and atmospheric cosmic-ray spectra with the BESS-TeV spectrometer. *Phys. Lett. B* **2004**, *594*, 35–46. [CrossRef]
20. Dmitriyeva, E.; Fedosimova, A.; Lebedev, I.; Temiraliev, A.; Abishev, M.; Kozhamkulov, T.; Mayorov, A.; Spitaleri, C. Determination of the primary energy using an ultrathin calorimeter. *J. Phys. G Nucl. Part. Phys.* **2020**, *47*, 035202. [CrossRef]
21. Allison, J.; Amako, K.; Apostolakis, J.; Arce, P.; Asai, M.; Aso, T.; Bagli, E.; Bagulya, A.; Banerjee, S.; Barrand, G.; et al. Recent developments in Geant4. *Nucl. Instrum. Methods Phys. Res. Sect. A Accel. Spectrometers Detect. Assoc. Equip.* **2016**, *835*, 186–225. [CrossRef]
22. Fedosimova, A.; Gaitinov, A.; Grushevskaya, E.; Lebedev, I. Study of the peculiarities of multiparticle production via event-by-event analysis in asymmetric nucleus-nucleus interactions. *EPJ Web. Conf.* **2017**, *145*, 19009. [CrossRef]
23. Fedosimova, A.; Gaitinov, A.S.; Lebedev, I.A.; Temiraliev, A. Study on initial geometry fluctuations via correlation of finite distributions of secondary particles in nucleus-nucleus interactions. *J. Phys. Conf. Ser.* **2016**, *668*, 012067. [CrossRef]
24. Kvochkina, T.N.; Lebedev, I.A.; Lebedeva, A.A. An analysis of high-energy interactions with large transverse momentum of secondary particles. *J. Phys. G Nucl. Part Phys.* **2000**, *26*, 35–41. [CrossRef]
25. Adamovich, M.I.; Andreeva, N.P.; Basova, E.S.; Bradnova, V.; Bubnov, V.I.; Chernyavsky, M.M.; Gaitinov, A.S.; Gulamov, K.G.; Haiduc, M.; Hasegan, D.; et al. Flow Effects in High-Energy Nucleus Collisions with Ag(Br) in Emulsion. *Phys. At. Nucl.* **2004**, *67*, 273–280. [CrossRef]
26. Adamovich, M.I.; Aggarwal, M.M.; Andreeva, N.P.; Badyal, S.K.; Bakich, A.M.; Basova, E.S.; Bhalla, K.B.; Bhasin, A.; Bhatia, V.S.; Bradnova, V.; et al. Factorial Moments of ^{28}Si Induced Interactions with Ag(Br) Nuclei. *APH N.S. Heavy Ion Phys.* **2001**, *13*, 213–221. [CrossRef]
27. Lipari, P. Universality in the longitudinal development of Cosmic Ray showers. *Nucl. Part. Phys. Proc.* **2016**, *279–281*, 111–117. [CrossRef]
28. Fedosimova, A.; Kharchevnikov, P.; Lebedev, I.; Temiraliev, A. Applying universality in the development of cascade processes for the research of high energy cosmic particles in space experiments. *EPJ Web. Conf.* **2017**, *145*, 10004. [CrossRef]
29. Tompakova, N.; Dmitriyeva, E.; Lebedev, I.; Serikkanov, A.; Grushevskaya, Y.; Mit', K.; Fedosimova, A. Influence of hydrogen plasma on SnO_2 thin films. *Mater. Today Proc.* **2020**, *25*, 83–87. [CrossRef]

Article

Measurement of Energy Spectrum and Elemental Composition of PeV Cosmic Rays: Open Problems and Prospects

Giuseppe Di Sciascio

INFN—Roma Tor Vergata, Department of Physics, University of Roma Tor Vergata, Viale della Ricerca Scientifica 1, I-00133 Roma, Italy; disciascio@roma2.infn.it

Abstract: Cosmic rays represent one of the most important energy transformation processes of the universe. They bring information about the surrounding universe, our galaxy, and very probably also the extragalactic space, at least at the highest observed energies. More than one century after their discovery, we have no definitive models yet about the origin, acceleration and propagation processes of the radiation. The main reason is that there are still significant discrepancies among the results obtained by different experiments located at ground level, probably due to unknown systematic uncertainties affecting the measurements. In this document, we will focus on the detection of galactic cosmic rays from ground with air shower arrays up to 10^{18} eV. The aim of this paper is to discuss the conflicting results in the 10^{15} eV energy range and the perspectives to clarify the origin of the so-called 'knee' in the all-particle energy spectrum, crucial to give a solid basis for models up to the end of the cosmic ray spectrum. We will provide elements useful to understand the basic techniques used in reconstructing primary particle characteristics (energy, mass, and arrival direction) from the ground, and to show why indirect measurements are difficult and results are still conflicting.

Keywords: cosmic ray physics; multi-messenger astrophysics; extensive air showers

1. Introduction

Cosmic rays (CRs) are the most outstanding example of accelerated particles and represent about 1% of the total mass of the universe [1]. The riddle of the origin of this radiation has been unsolved for more than a century. The study of CRs is based on two complementary approaches [2]:

(1) Measurement of energy spectrum, elemental composition and anisotropy in the CR arrival direction distribution, the three basic parameters crucial for understanding the origin, acceleration, and propagation of radiation.

(2) Search of their sources through the observation of neutral radiation (photons and neutrinos), which points back to the emitting sources not being affected by the magnetic fields, in a multi-messenger approach. We note that, however, photons and neutrinos do not necessarily point back to their sources (see, for example, the the Ref. [3]).

In Figure 1, the primary CR all-particle energy spectrum (namely, the number of nuclei as a function of total energy) is shown. The spectrum exceeds 10^{20} eV, showing a few basic characteristics [2]:

(a) A power-law behaviour $\sim E^{-2.7}$ up to the so-called *"knee"*, a small downwards bend around a few PeV (1 PeV = 10^{15} eV);

(b) a power-law behaviour $\sim E^{-3.1}$ beyond the knee, with a downwards bend near 10^{17} eV, sometimes referred to as the *"second knee"*;

(c) a transition back to a power-law $\sim E^{-2.7}$ (the so-called *"ankle"*) around $10^{18.7}$ eV;

(d) a cutoff, probably due to extra-galactic CR interactions with the Cosmic Microwave Background (CMB), around $10^{19.7}$ eV (the Greisen-Zatsepin-Kuzmin effect).

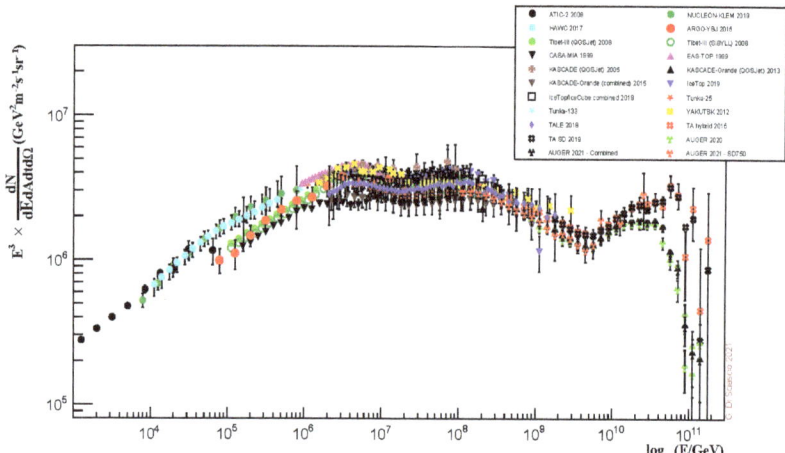

Figure 1. All–particle energy spectrum of primary cosmic rays, the flux is multiplied by E^3. Results from direct and indirect experiments updated to the year 2021 are shown.

Despite the differences in flux, emphasized by multiplying the differential spectrum by E^3, all the measurements of the all-particle energy spectrum are in fair agreement when taking into account the statistical, systematic and energy scale uncertainties. Nevertheless, uncertainties affecting flux measurements could be underestimated for a number of reasons discussed in this document. In a conservative approach, the spread of different results provides a more realistic estimate of the uncertainty.

All the observed features are believed to carry fundamental information that sheds light on the key questions of the origin, acceleration and propagation of CRs. However, from the all-particle results alone, it is not possible to understand the origin of different features. All models concerning sources, acceleration and propagation of the primary flux, differ considerably for what concerns expected elemental composition as a function of the energy. A measurement of the chemical composition is therefore crucial to disentangle between different hypotheses.

The main structure is the *"knee"* observed for the first time by R.W. Williams in 1948 in the experiment which first located individual shower cores from symmetry of the fired detectors [4,5]. The knee as a feature connected to the end of the Galactic CR flux was first suggested in 1959 by Kulikov and Khristiansen [6]. They speculated that particles above 10^{16} eV may have a *"metagalactic origin"*. Consequently, the observed spectrum is a superposition of the spectra of particles of galactic and metagalactic origin. In 1962, Miura and Hasegawa [7] reported the first observation of two spectral kinks (in both N_e and N_μ spectra) correlating them to a steepening of the primary energy spectrum.

All experiments observed the knee at about 4×10^{15} eV but a general consensus about the chemical component responsible for such a feature does not exist yet because experimental results are still conflicting, as will be discussed in Section 5. Determining elemental composition in the knee energy region is crucial to understand where Galactic CR spectrum ends and to give a solid basis to CR models up to the highest observed energies. The maximum energy at which the various nuclei are accelerated should be subject to a rigidity cutoff, as proposed originally by Peters [8]. Protons will cutoff first, followed by other nuclei according to the relation

$$E_{\max}(Z) = Z \times E_{\max}(Z=1) \quad (1)$$

If the dominant primary mass of the knee is light (protons and helium), then, according to this scheme, the Galactic CR spectrum is expected to end around 10^{17} eV with iron. The sum of the fluxes of all elements, with their individual knees at energies proportional to the nuclear charge, makes up the CR all-particle spectrum shown in Figure 1. With increasing energies, not only does the spectrum become steeper due to such cutoffs, but also heavier. In this scenario, the knee would represent the end of the spectrum of CR accelerated by SNRs in the galaxy.

Indeed, it is widely believed that the bulk of CRs up to about 10^{17} eV are galactic, produced and accelerated by the shock waves of SuperNova Remnants (SNR) expanding shells [9], and that the transition to extra-galactic CRs occurs somewhere between 10^{17} and 10^{19} eV. The experimental results, however, do not demonstrate the capability of SNRs to produce the power needed to sustain the population of galactic CRs and to accelerate particles up to the knee, and beyond. Indeed, to accelerate protons up to the PeV energy domain, a significant amplification of the magnetic field at the shock is required, but this process is problematic [10].

Unlike neutrinos that are produced only in hadronic interactions of CRs, the question whether the observed γ-rays are produced by the decay of π^0 from CR interactions ('*hadronic*' mechanism), or by a population of relativistic electrons via Inverse Compton scattering or bremsstrahlung ('*leptonic*' mechanism), still needs a conclusive answer. In a hadronic interaction, the secondary photons have, on average, an energy factor of 10, lower than the primary proton. Therefore, the quest for CR sources to be able to accelerate particles up to the PeV range in a multi-messenger approach requires the observation of the γ-ray sky above 100 TeV. However, the first results reported by the LHAASO experiment [11,12], that is, the observation of a number of gamma sources emitting photons beyond 500 TeV, show that SNRs are likely not the main sources of PeV CRs in our galaxy. In fact, none of the 12 observed ultra-high energy gamma sources can be clearly described with hadronic mechanisms operating in SNRs. We note that the highest photon emission at 1.4 PeV comes from a system of massive stars in the Cygnus Region, the so-called '*Cygnus Cocoon*', a possible factory of fresh CRs, as suggested by other experiments [13,14].

In this note, we will focus on galactic CRs in the PeV energy range detected from ground with air shower arrays. This is not a place for a complete review of CR physics and models (for which we recommend, for instance, [2,15–19] and the references therein), but only to provide elements useful to understand the main techniques used in reconstructing primary particle characteristics from the ground with particle arrays, and to show why indirect measurements are difficult and the results are still conflicting.

In the next section, we will introduce the detection techniques. In Section 3 we will describe the main characteristics of Extensive Air Showers to understand how different observables measured by arrays are related to the properties of the primary CRs. In Section 4, we will discuss the general scheme of the air shower array analysis. In Section 5 the experimental results in the 10^{14}–10^{18} eV energy range are summarized. The prospects for new measurements in the knee region are introduced in Section 6.

2. Detection Techniques

We can divide the experimental methods adopted to measure fluxes and elemental composition of CRs into two categories: '*direct*' and '*indirect*' measurements. Generally speaking, for all particle types:

- the higher the energy, the lower the flux;
- the lower the flux, the larger the required detector area.

The direct measurements, in principle, detect and directly identify the primary particles with detectors outside the atmosphere (on board of stratospheric balloons or satellites), since the atmosphere behaves as a shield (see below). Since the CR flux rapidly decreases with increasing energy and the size of detectors is constrained by the weight that can be carried in flight, their '*aperture*' (i.e., the acceptance measured in m$^2 \cdot$sr) is small and determines a maximum energy (of the order of a few hundred TeV/nucleon), at which a statistically significant detection is possible. In fact, the number of detected events is given by the CR flux times the detection area times the total observation time. Therefore, the detection area limits the smallest measurable flux. In addition, the limited volume of the detectors makes the containment of showers induced by high-energy nuclei difficult, thus limiting the energy resolution of the instruments in direct measurements.

At higher energies, the flux is so low (about 1 particle/m^2/year around 10^{15} eV) that the only chance is to have earth-based detectors of large area, operating for long times. In that case, the atmosphere is considered as a target, and we study the primary properties in an *'indirect'* way, through the measurement of secondary particles produced in the interaction of the primary particle with the nuclei of the atmosphere, the so-called *'Extensive Air Shower'* (EAS).

Approaching the hundred TeV energy region, even in space-borne experiments, the energy assignment is indirect since it is generally based on the energy deposition of particles produced in the interaction of primaries in the detector itself. The reconstruction of the total energy is then obtained by comparison with some model prediction, and therefore, at least in that region, the boundary line between 'direct' and 'indirect' experiments is more uncertain. In fact, important results obtained by 'direct' methods are conflicting due to some still unknown systematic uncertainties probably related to the interaction model used to assign the energy. A neutrino energy-dependent component must be estimated via Monte Carlo simulations, an evaluation which adds some additional model dependency for 'indirect' measurements.

At the ground, the study of CRs is based on the reconstruction and interpretation of EAS observables in the different components, electromagnetic (e.m.), muonic and hadronic, Cherenkov photons, nitrogen fluorescence, radio emission. Therefore, different detectors must be used to detect different observables.

Two different approaches are exploited:

- Arrays, to sample the shower tail particles reaching the ground. In High Energy Particle language, a shower array is a *"Tail Catcher Sampling Calorimeter"*. The atmosphere is the absorber and the detectors at ground are the device to measure a (poor) calorimetric signal. Arrays are wide field of view detectors able to observe most of the overhead sky with a duty cycle of ~100%. Measurements are limited by large shower-to-shower fluctuations.
- Telescopes, to detect Cherenkov photons or nitrogen fluorescence and observe the EAS longitudinal profile. The atmosphere acts as a *"Homogeneous Calorimeter"*. The duty cycle is low (~10–15%) because telescopes can be operated only during clear moonless nights and the field of view is small (a few degrees). On the contrary, pointing capability and energy resolution are excellent.

Shower arrays are made by a large number of detectors (scintillators, Resistive Plate Chambers (RPCs) or water Cherenkov tanks, for example) distributed in a regular grid over very large areas, of an order of 10^4–10^5 m^2 (see Figure 2). The shower *"size"*, the total number of charged particles, and the shower arrival direction are the two key parameters reconstructed by all arrays. The majority of EAS arrays do not distinguish between the charged particles. From the measurement of the particle densities on the fired detectors of the array it is possible to determine the shower core position, that is, the point where the shower axis intersects the detection plane, and, via a Lateral Density Function (LDF), reconstruct the size of the shower. The LDF is of phenomenological nature, determined via Monte Carlo simulations for the particular experimental set-up [20]. The direction of the incoming primary particle is reconstructed with a *'time of flight'* method making use of the relative times at which the individual detection units are fired by the shower front [20].

Figure 2. Example of a typical air shower array (Tibet ASγ experiment located at the YangBaJing Cosmic Ray Observatory in Tibet (P.R. China) 4300 m asl).

On general grounds, the instrumented area A determines the rate of high energy events recorded, that is, the maximum energy via limited statistics. The grid distance d determines the low energy threshold (small energy showers are lost in the gap between detectors) and the quality of the shower sampling. The particular kind of detector (scintillator, RPC, water tank) determines the detail of measurement (efficiency, resolution, energy threshold, quality) and impact on the cost per detector C_d. In principle, best physics requires large area A, small distance d and high quality of the sampling. However, the cost of an array increases with $C_d \cdot A/d^2$, therefore a compromise is always needed. This is one of the reason why the typical total sensitive area of a classical array is less than 1% of the total enclosed area. This results in a high degree of uncertainty in the reconstruction due to sampling fluctuations which add to the shower fluctuations.

The experiments devoted to study the PeV energy range have been operated at different altitudes, ranging from the extreme altitude (5200 m asl) of BASJE-MAS [21] to the sea level of KASCADE [22–24].

In Tables 1 and 2, the characteristics of air shower arrays operated in the last two decades to study Galactic CR physics from ground are summarized. The atmospheric depths of the arrays, the main detectors used, the energy range investigated, the sensitive areas of e.m. and muon detectors, the instrumented areas and the coverage (i.e., the ratio between sensitive and instrumented areas) are reported. The depth in atmosphere is crucial to fix the energy threshold, the energy resolution, the impact of shower-to-shower fluctuations, then the sensitivity to elemental composition.

Table 1. Characteristics of different air shower arrays.

Experiment	g/cm²	Detector	ΔE (eV)	e.m. Sens. Area (m²)	Instr. Area (m²)	Coverage
ARGO-YBJ [20]	606	RPC/hybrid with wide-FoV Č Tel.	3×10^{11}–10^{16}	6700	11,000	0.93 (carpet)
BASJE-MAS [21]	550	scint./muon	6×10^{12}–3.5×10^{16}		10^4	
TIBET ASγ [25]	606	scint./burst det.	5×10^{13}–10^{17}	380	3.7×10^4	10^{-2}
CASA-MIA [26]	860	scint./muon	10^{14}–3.5×10^{16}	1.6×10^3	2.3×10^5	7×10^{-3}
KASCADE [22]	1020	scint./mu/had	2×10^{15}–10^{17}	5×10^2	4×10^4	1.2×10^{-2}
KASCADE-Grande [27]	1020	scint./mu/had	10^{16}–10^{18}	370	5×10^5	7×10^{-4}
Tunka [28]	900	open Č det.	3×10^{15}–3×10^{18}	—	10^6	—
IceTop [29]	680	ice Č det.	10^{16}–10^{18}	4.2×10^2	10^6	4×10^{-4}
LHAASO [11]	600	Water Č scint./mu/had wide-FoV Č Tel.	10^{12}–10^{17}	5.2×10^3	1.3×10^6	4×10^{-3}

Table 2. Characteristics of different muon detectors operated in some shower arrays.

Experiment	Altitude (m)	μ Sensitive Area (m²)	Instrumented Area (m²)	Coverage
LHAASO	4410	4.2×10^4	10^6	4.4×10^{-2}
TIBET ASγ	4300	4.5×10^3	3.7×10^4	1.2×10^{-1}
KASCADE	110	6×10^2	4×10^4	1.5×10^{-2}
CASA-MIA	1450	2.5×10^3	2.3×10^5	1.1×10^{-2}

Generally speaking, near the depth of the maximum of the shower development, the number of secondary charged particles is almost independent of the mass of the primary particle, and the shower fluctuations are at minimum. For the knee energy region, this depth corresponds to ≈5000 m asl. Therefore, these extreme altitudes are suitable to have good energy resolution, to reconstruct the primary energy in a mass-independent way and to study the shower core region in great detail, where the hadronic component feeds the e.m. one deep in the atmosphere. As demonstrated by the ARGO-YBJ [30] and Tibet ASγ [31] experiments, observables related to the shower core properties are almost independent on the details of hadronic interaction models. At high altitudes, due the low energy threshold (≈TeV), it is possible to cross-check the fluxes with direct measurements on a wide energy range (ARGO-YBJ in the 5–250 TeV range). This cross-calibration is important due to the conflicting results obtained not only by ground-based detectors, but also by direct experiments. In addition, the absolute energy scale can be calibrated at a level of 10%, exploiting the so-called *"Moon Shadow"* technique [20].

On the other hand, experiments located deep in the atmosphere enhance the differences in the longitudinal development of EAS of different primary masses, as the shower is sampled well beyond its maximum. Therefore, the ratio N_e/N_μ is, in principle, more suitable for elemental composition studies. However, shower fluctuations are much larger, making it difficult to interpret the data. In addition, the reconstruction of the energy is typically strongly model-dependent because 'a priori' assumptions on the primary composition are needed, and the calibration of the absolute energy scale is one of the major open issues.

The great variety of layouts, observables, and reconstruction procedures to infer the elemental composition is at the origin, in part, of the conflicting results reported by different ground-based experiments. Arrays focused on the investigation of the knee region operated so far are also characterized by a limited size of the instrumented area. They collected limited statistics above 10^{16} eV, and were, therefore, unable to give a conclusive answer to the origin of the knee. The poor sensitivity to elemental composition, due to the small statistics, prevents discrimination against different mass groups, and only general trends can be investigated in terms of the evolution of $\langle \ln A \rangle$ or of "*light*" and "*heavy*" components with energy.

3. Extensive Air Showers: The Heitler-Matthews Model

A general idea of the main characteristics of EAS and of how different nuclei produce showers with different properties can be obtained from some relatively simple arguments, as suggested by Heitler [32] and Matthews [33]. This toy model is useful to show how different observables depend on the primary mass and energy, and why certain techniques have historically been used to study elemental composition or to reconstruct the energy spectrum. Nevertheless, detailed Monte Carlo simulations must be used to describe quantitatively all the characteristics of these random processes, with particular care to the role of shower fluctuations.

In a nutshell, the collision of a primary CR with a nucleus of the atmosphere produces one large nuclear fragment and many charged and neutral pions (with a smaller number of kaons) (Figure 3) [34]. A significant fraction of the total energy is carried away by a single "*leading*" particle. This energy is unavailable immediately for new particle production. Roughly speaking, half of the energy of the primary particle is transferred to the nuclear fragment and the other half is taken by the pions (and kaons). The fraction of energy transferred to the new shower particles is referred as *inelasticity*. Accurate description of the leading particles is crucial because these high-energy nucleons feed energy deeper into the atmosphere. Approximately equal number of positive, negative and neutral pions are produced. The e.m. component, the most intense of an EAS, is produced by the photons coming from the decay of the neutral pions. At each interaction before the charged pions decay, nearly a third of the hadronic component energy is released into the e.m. one.

As the number of particles increases, the energy per particle decreases. They will also scatter, losing energy, and many will range-out. Thus, the number of particles (or, with less ambiguities in the definition, the quantity of energy transferred to secondaries and eventually released into the atmosphere) will reach a maximum at some depth X_{max} which is a function of energy, of the nature of the primary particle and of the details of the interactions of the particles in the cascade. After that, the energy/particle is so degraded (will be below some "*critical energy*") that energy losses dominate over particle multiplication process, and the shower "size" will decrease as a function of depth: it grows 'old'. Once the pions have reached an energy which is low enough, they will decay into muons and neutrinos ($\pi^+ \to \mu^+ \nu_\mu$ or $\pi^- \to \mu^- \bar{\nu}_\mu$). The resulting muons propagate unimpeded to the ground. The muon cascade grows and maximizes, but the decay is slower as a consequence of the relative stability of the muon and small energy losses by ionization and pair production.

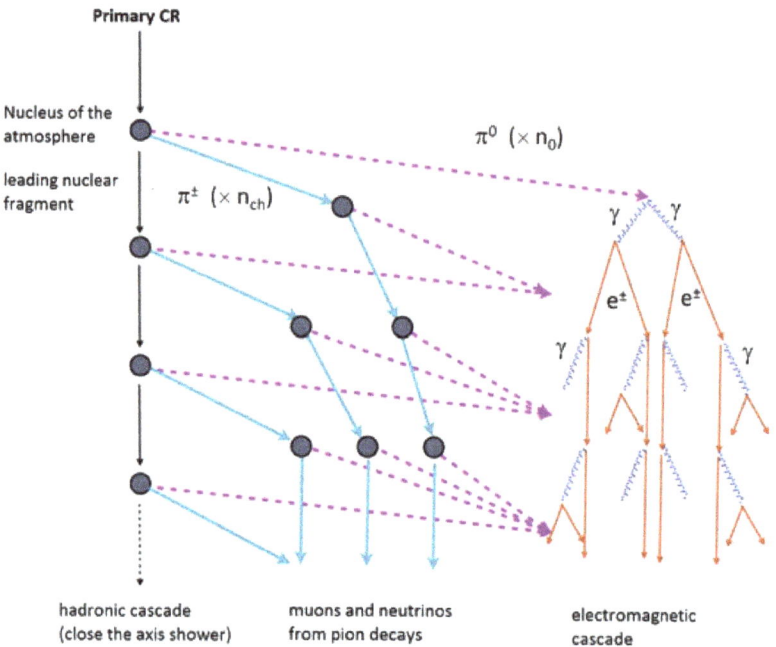

Figure 3. Schematic evolution of cascades initiated by a CR particle. At each step, roughly 1/3 of the energy is transferred from the hadronic cascade to the e.m. one. Figure taken from [34].

These are the most common processes, but not at all the only ones. As an example, successive hadronic interactions of the primary CR, interactions/decays of kaons and muon decays, multiple scattering and production angles must also be considered (see, for example, the the Ref. [16]). Only detailed simulations with Monte Carlo methods are able to describe all the characteristics of these random processes.

Historically, one of the main problems in analyzing data from shower arrays was related to the fact that each experiment used its own simulation of shower development and detectors. This made difficult the comparison of the results and the understanding of their differences. Starting in the 1990s, all experiments began to use the same Monte Carlo simulation code CORSIKA [35], a framework containing different hadronic interaction models to describe the shower development in the atmosphere, and the software GEANT [36] to simulate the detectors operated in the arrays. Over the years, other simulation codes have been developed, in particular to describe the development of showers at ultra-high energies, such as AIRES [37]. The main characteristics of hadronic interactions that are relevant for EAS physics are: cross-sections (p–air, π–air, N–air), inelasticity of the collisions, multiplicity/composition of secondaries, transverse momentum distribution, fraction of diffractive dissociation.

New data coming from the LHC (at an energy $E_{lab} \sim 10^{15}$ eV) allowed to improve the models even if some points remain critical. In fact, the situation is much worse than it may appear from energy considerations. Measurements at colliders are limited to an angular region that excludes the beam pipe (the so-called *'central region'*), and therefore a very large majority of the high-energy particles that are emitted at small angles (in the so-called *'forward region'*) are unobservable. In EAS physics the forward region is the most relevant because the high-energy particles feed energy in the shower down in the atmosphere. Therefore, models tuned to accelerator measurement in the central region are extrapolated to describe the interactions of CRs.

Nevertheless, this simple toy model predicts the basic features of EAS development. In the following, the e.m. and hadronic processes will be described separately in more detail.

3.1. Electromagnetic Showers

The main features of an e.m. shower profiles can be described within the simple Heitler's toy model of particle cascades [32]. Let us suppose that a particle (electron, positron or photon) with energy E_0 splits its energy equally into two particles after traveling a radiation length X_0 in air, and let this process be repeated by the secondaries (see Figure 4).

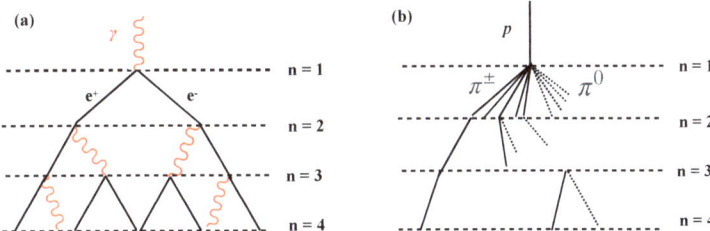

Figure 4. Schematic view of an e.m. cascades (**a**), and of a hadronic shower (**b**). In the hadron shower, dashed lines show π^0 which do not re-interact but decay, producing e.m. sub-showers.

Let X describe the depth in the atmosphere and define the depth at which the average CR starts interactions with the atmosphere to be $X = 0$ g/cm^2. After n radiation lengths, we obtain a particle cascade which has evolved into $N = 2^n$ particles of equal energy $E = E_0/N$. Multiplication stops when the energies of the particles are too low for pair production or bremsstrahlung. This energy is the critical energy ε_c^{em} in the air ($\approx$80 MeV, below which the collisional energy losses are dominant).

The maximum number of particles N_{max} is reached at this moment, when all particles have the same energy ε_c^{em}, $E_0 = \varepsilon_c^{em} \cdot N_{max}$. The depth X_{max} at which the shower reaches the maximum size is $X_{max} = n_{max} \cdot X_0$, where n_{max} is the number of radiation lengths required for the primary energy to be reduced to ε_c^{em}.

Since $N_{max} = 2^{n_{max}}$, we have

$$n_{max} = \ln\left(\frac{E_0}{\varepsilon_c^{em}}\right) \cdot \frac{1}{\ln 2} \quad (2)$$

so that

$$X_{max}^{em} = \frac{X_0}{\ln 2} \cdot \ln\left(\frac{E_0}{\varepsilon_c^{em}}\right). \quad (3)$$

Finally, it is interesting to estimate the *elongation rate* Λ, that is, the rate of increase of X_{max} with the primary energy. From the relation (3), we have

$$\Lambda^{em} = \frac{d\,X_{max}}{d\,\log_{10} E_0} = 2.3 \cdot X_0 = 85 \text{ g/cm}^2 \quad \text{per decade of energy.} \quad (4)$$

This simple model predicts two basic features of e.m. shower development:

- N_{max} increases proportional to the primary energy E_0, $N_{max} = \frac{E_0}{\varepsilon_c^{em}}$.
- X_{max} increases logarithmically with primary energy, at a rate of 85 g/cm^2 per decade of energy.

3.2. Hadronic Showers

Air showers initiated by protons have been modeled by different authors (see, for example, the the Ref. [38–40]) following the Matthews approach [33], similar to the Heitler one. The main differences with the e.m. cascades are

1. in the hadronic interactions a large number of secondary particles are produced. At high energy, the total multiplicity of particles per collision, N_{tot}, reaches values of several tens with the consequence that although the hadronic interaction lengths are larger than the e.m. radiation one (at PeV energies $\lambda_{p-air} \simeq 80$ g cm^{-2}), the showers develop faster than in the e.m. case;
2. in a hadronic interaction only a fraction of the energy E_0 is available for secondary particle production. A single leading particle, the highest energy secondary produced in the interaction, carries a fraction $(1-\kappa)E_0$ deep in the atmosphere, where κ is the so-called *inelasticity*. Therefore, a fraction $\frac{2}{3}\kappa E_0$ is used to produce N_{ch} charged pions, and a fraction $\frac{1}{3}\kappa E_0$ goes via neutral pions into the e.m. component;
3. the critical energy ε_c^π is defined as the energy at which the decay and the hadronic interaction probabilities are equal and further particle production by $\pi^\pm$ ceases.

According to the Ref. [33], constant values $N_{ch} = 10$, corresponding to an energy of about 100 GeV, and $\varepsilon_c^\pi = 20$ GeV are adopted in the following.

Protons travel one interaction length and interact producing N_{tot} pions, all having equal energies, N_{ch} are charged and $\frac{N_{tot}}{3} = \frac{1}{2} \cdot N_{ch}$ neutral, which immediately decay into photons, initiating e.m. showers. As for the e.m. cascade, we assume equal division of energy during particle production.

In turn, the charged pions can decay in muons and neutrinos and hence, as long as their decay length remains larger than their interaction length, they will re-interact rather than decay. This happens for $\gamma c \tau_\pi > \lambda_{\pi-air}/\rho_{air}$, with the Lorentz factor $\gamma = E_\pi/m_\pi$, the charged pion lifetime $\tau_\pi \simeq 26$ ns and $\lambda_{\pi-air} \simeq 1.5\lambda_{p-air} \simeq 120$ g cm^{-2} (since the πp cross-section is about 2/3 the pp cross-section). This implies that pions will re-interact as long as their energy satisfies $E > E_d \simeq 100$ GeV$(10^{-4}$ g cm$^{-3}/\rho_{air})$. Hence, at the heights above 10 km, where the initial development of the shower takes place, $\pi^\pm$ will re-interact for energies greater than $\sim$20–30 GeV [40].

After n interactions, the $N_\pi = (N_{ch})^n$ charged pions produced carry a total energy of $(\frac{2}{3})^n \cdot E_0$. The energy per charged pion after n interactions is then $E_{\pi^\pm} = \frac{E_0}{(3/2N_{ch})^n}$. The remainder of the primary energy goes into the e.m. component from π^0 decays

$$E_{em} \simeq E_0 \left[1 - \left(\frac{2}{3}\right)^n\right]. \quad (5)$$

After only six interactions, about 90% of the initial energy is transferred to the e.m. component of the shower, with the remaining 10% being essentially the muons and neutrinos from the charged pion decays. As a consequence, most of the energy of an air shower can be observed in its e.m. component. This is the so-called *calorimetric energy* which allows to estimate the primary energy with good accuracy to detectors able to observe the longitudinal air shower development.

Assuming that at ε_c^π, all pions decay, the number of muons is $N_\mu = N_{\pi^\pm} = (N_{ch})^{n_c}$, where n_c is the number of interaction lengths required for the charged pion's interaction length to exceed its decay length

$$n_c = \frac{\ln(E_0/\varepsilon_c^\pi)}{\ln(\frac{3}{2}N_{ch})} = 0.85 \lg\left(\frac{E_0}{\varepsilon_c^\pi}\right). \quad (6)$$

Thus, the total energy is divided into two channels, hadronic and electromagnetic

$$E_0 = E_{em} + E_h = \varepsilon_c^{em} \cdot N_e + \varepsilon_c^\pi \cdot N_\mu. \quad (7)$$

This equation represents *energy conservation*, apart from a fraction of a few percent of the primary energy spent in the neutrino component. The relative magnitude of the contribution from N_μ and N_e does not depend on the details of the model, but only on the respective critical energies, the energy scales at which e.m. and hadronic multiplication ceases. An important conclusion of this description of the hadronic cascades is that the energy is given by a linear combination of muon and electron sizes. This result is insensitive to fluctuations in the division of energy between the hadronic and e.m. channels and independent on the mass of the primary particle.

The number of muons is given by

$$\ln N_\mu = \ln N_{\pi^\pm} = n_c \ln N_{ch} = \frac{\ln(E_0/\varepsilon_c^\pi)}{\ln(3/2N_{ch})} \cdot \ln(N_{ch}) = \beta \cdot \ln\left(\frac{E_0}{\varepsilon_c^\pi}\right). \quad (8)$$

Following [33], we can estimate $\beta = \frac{\ln(N_{ch})}{\ln(3/2N_{ch})} = 0.85$ for E_0 in the range 10^{14}–10^{17} eV, obtaining

$$N_\mu = \left(\frac{E_0}{\varepsilon_c^\pi}\right)^\beta = \left(\frac{E_0}{\varepsilon_c^\pi}\right)^{0.85} \sim 9900 \left(\frac{E_0}{10^{15}\text{eV}}\right)^{0.85}. \quad (9)$$

Including inelasticity in the Heitler model [33] changes the parameter β

$$\beta = \frac{\ln(N_{ch})}{\ln(3/2N_{ch})} \to \frac{\ln[1 + N_{ch}]}{\ln\left[(1+N_{ch})/(1-\frac{1}{3}\kappa)\right]} \approx 1 - \frac{\kappa}{3\ln(N_{ch})} = 1 - 0.14\kappa. \quad (10)$$

The elasticity for the most energetic meson in pion–air interactions yields $(1 - \kappa)$ between 0.26 and 0.32, resulting in $\beta = 0.90$.

The electronic size can be calculated by inserting the expression (9) for the muon size in the energy conservation relation (7)

$$\frac{E_{em}}{E_0} = \frac{E_0 - N_\mu \varepsilon_c^\pi}{E_0} = 1 - \left(\frac{E_0}{\varepsilon_c^\pi}\right)^{\beta-1}. \quad (11)$$

The e.m. fraction is 66% at $E_0 = 10^{15}$ eV, increasing to 83% at 10^{18} eV for proton-induced showers.

Therefore, the number of electrons at a maximum shower for proton-induced showers is

$$N_e = \frac{E_{em}}{\varepsilon_c^{em}} = \frac{E_0}{\varepsilon_c^{em}} - \frac{\varepsilon_c^\pi}{\varepsilon_c^{em}}\left(\frac{E_0}{\varepsilon_c^\pi}\right)^\beta \approx \frac{E_0}{\varepsilon_c^{em}} = N_{e|max}^p. \quad (12)$$

The approximation is justified at high energies when the fraction of energy transferred to muons is small [39].

In the framework of the *superposition model*, each nucleus is taken to be equal to A individual single nucleons, each with energy E_0/A and each acting independently. The shower resulting from the interaction of the primary nucleus A can be treated as the sum of A proton-induced independent showers all starting at the same point. Thus, while a proton creates one shower with energy E_0, an iron nucleus of the same total energy is expected to create the equivalent of 56 proton showers, each with reduced energy $(E_0/56)$. The average properties of showers are well reproduced by this model, though the fluctuations are clearly underestimated and can be studied only with detailed Monte Carlo simulations of the intra-nuclear cascade. The superposition of A independent showers naturally explains why the shower-to-shower fluctuations are smaller for shower initiated by nuclei as compared to proton showers.

By substituting the lower primary energy (E_0/A) into the previous expressions and summing A such showers, we obtain the following relations for the number of electrons and muons in a shower induced by a nucleus A:

$$N_{e|max}^A = A\left(\frac{E_0/A}{\varepsilon_c^{em}}\right) = N_{e|max}^p \quad (13)$$

$$N_{\mu|max}^A = \left(\frac{E_0}{\varepsilon_c^\pi}\right)^\beta A^{1-\beta} = N_{\mu|max}^p A^{1-\beta} \approx 1.69 \cdot 10^4 \cdot A^{0.10}\left(\frac{E_0}{1\text{ PeV}}\right)^{0.90}. \quad (14)$$

From these relations valid *at shower maximum* follows:

1. The number of electrons is equal for all primary masses A, that is, is independent of the composition. Therefore, *the shower size $N^A_{e|max}$ can be used as an estimator of the energy;*

2. The number of muons $N^A_{\mu|max}$ increases with the mass of the primary particle with $A^{1-\beta} \sim A^{0.1}$. Accordingly, iron-induced showers contain about 1.5 times as many muons as proton showers with the same energy. In fact, in a shower induced by a nucleus A, due to the smaller energy per nucleon (E_0/A), the secondary pions are less energetic. This favours a pion decay as well as an interaction of heavier nuclei higher in atmosphere, where the air density is smaller. *The number of muons can be used to infer the mass of the primary particle.* Moreover, the evolution of the muon number with energy, $dN_\mu/d \ln E$, is a good tracer of changes in the primary composition. In fact, a constant composition gives $dN_\mu/d \ln E = \beta$ and any departure from that behavior can be interpreted as a change of the average mass of the primaries, in a similar way as with the elongation rate of the longitudinal development.

3. The muon size grows with primary energy more slowly than proportionally, $\beta \sim 0.90$.

A large number of ground-based arrays studying the knee energy region are located deep in the atmosphere and do not sample the number of electrons at shower maximum. Therefore, the experimental situation is not ideal because the size, used to recover the energy of the primary particle, is mass-dependent, as discussed in Section 4. Only experiments located at extreme altitude (above 4000 m asl) observe the electrons in the shower maximum region for near-vertical showers with an energy in the PeV range.

Deeper in the atmosphere, arrays measure only the attenuated size

$$N_{e|ground} \approx N_{e|max} \cdot \exp\left(-\frac{\Delta X}{\Lambda}\right) \quad (15)$$

where ΔX is the distance of the shower maximum from the ground and $\Lambda \approx 60$ g/cm^2 is the attenuation length of the electron size after the shower maximum. Since heavy nuclei reach the maximum of longitudinal development at smaller depths than light ones, on the ground we have a larger electron number for air showers initiated by light particles. This implies that, due to the steeply falling CR spectrum, showers of equal $\ln N_e$ are enriched in light elements.

3.3. Longitudinal Development

The longitudinal development of a hadronic shower is dominated by the parallel e.m. sub-showers produced by the π^0 decays in the first interaction, at an atmospheric depth $X^* = \lambda_{p-air} \cdot \ln 2 \approx 55$ g/cm^2. In a good approximation following cascades can be neglected. The energy of the single photon is $E_\gamma = \frac{E_{\pi^0}}{2} = \frac{E_0}{3} \frac{2}{N_{ch}} = \frac{E_0}{3N_{ch}}$.

From Equation (3), we have

$$X^p_{max} = X^* + X_0 \cdot \ln\left(\frac{E_0}{3N_{ch} \cdot \epsilon^{em}_c}\right)$$
$$= X^* + X^{em}_{max} - X_0 \cdot \ln(3 N_{ch}) \quad \text{g/cm}^2$$

where X^{em}_{max} is the atmospheric depth of the maximum of γ-induced showers with E_0 primary energy and N_{ch} is the multiplicity of charged pions in the first interaction. The elongation rate for showers induced by protons is then

$$\Lambda^p = \Lambda^\gamma + \frac{d}{d \log_{10} E_0}\left[X^* - X_0 \cdot \ln(3 N_{ch})\right] = 58 \quad \text{g/cm}^2 \text{ per decade}, \quad (16)$$

reduced from the elongation rate for purely e.m. showers. This estimation verifies Linsley's elongation rate theorem [41], which points out that e.m. showers represent an upper limit to the elongation rate of the hadronic showers. The shower maximum is expected to be influenced by the elasticity of the first interaction, $(1-\kappa) = E_{lead}/E_0$, where E_{lead} is the energy of the leading particle. For interactions with $(1-\kappa) > 0.5$ most of the primary energy will be transferred deeper into the atmosphere and correspondingly the shower maximum will be deeper.

The extrapolation to a primary particle with mass A with the superposition model yields

$$X_{max}^A = X_{max}^p - X_0 \cdot \ln A \qquad (17)$$

Detectors able to observe the longitudinal air shower development can estimate the primary energy with good accuracy measuring the so-called *calorimetric energy*, that is, the energy of the e.m. component. With this estimator of the energy of the primary particle, the orthogonal variable sensitive to its primary mass is the depth of the shower maximum in terms of the number of particles, X_{max}.

Therefore:

- X_{max} is smaller for heavier nuclei (logarithmic dependence on A)
- X_{max} is the same for same E_0/A but different E_0. As a consequence, the proton-induced showers result, on average, in a larger number of particles at the observation level compared to iron-induced events. However, the shower-to-shower fluctuations are as large as the shift of X_{max} between proton and iron thus limiting an event-by-event assignement of a primary mass.

Despite the simplicity and the approximations of the toy model, the main characteristics of the EAS development are quite well reproduced. Obviously, a detailed description of the cascade, in particular for what concern the role of fluctuations, can be provided only by detailed Monte Carlo simulations.

3.4. Energy and Mass

The relevance of muon measurements to the question of the primary composition has been first remarked by the Institute for Nuclear Studies (INS) group in Tokyo [42]. They were the first group to point out the key information that the mass of the primary particle could be derived from a study of plots of muon versus electron number.

Due to the intuitive relation between shower to shower fluctuations and primary mass, the study of fluctuations in the muon number distributions was historically the first method employed to study the primary CR mass composition [42–44]. The narrowing of the distribution of N_μ/N_e was considered to be due partly to the change in the composition of primary particles with energy [43].

On general grounds, the elemental composition can be investigated if the total size and the muon component depend differently from the primary energy. If we assume that their dependences from the energy of a primary proton E_0 can be described as

$$N_e \propto E_0^{\beta_e}, \qquad N_\mu \propto E_0^{\beta_\mu}, \qquad (18)$$

for a nucleus A, we have

$$N_e \propto \left(\frac{E_0}{A}\right)^{\beta_e} \cdot A, \qquad N_\mu \propto \left(\frac{E_0}{A}\right)^{\beta_\mu} \cdot A, \qquad (19)$$

with a mass-number dependency of the type $1-\beta_e$ and $1-\beta_\mu$, respectively. The relation N_μ/N_e can be easily deduced

$$N_\mu \propto N_e^{\beta_\mu/\beta_e} A^{1-(\beta_\mu/\beta_e)}. \qquad (20)$$

In 1962, Linsley, Scarsi, and Rossi working at the MIT Volcano Ranch Station observed, for the first time, a muon/electron correlation: $N_\mu \sim A^{1-\alpha} \cdot (N_e)^\alpha$, thus establishing that the muon size is a mass-sensitive observable [45].

The Equation (13) can be transformed to obtain the energy E_0 to be introduced in the relation (14) to obtain $\beta_\mu/\beta_e \sim 0.86$. The muon size for a given mass A as a function of the total size N_e is then

$$N_\mu \propto N_e^{0.86} A^{0.14}. \tag{21}$$

In a similar way, we can obtain the muon size as a function of the total size for a given primary energy E_0. The Equation (13) is transformed to obtain the mass A which in turn is introduced in the relation (14)

$$N_\mu \propto \left(\frac{E_0}{1\ \text{PeV}}\right)^{3.17} N_e^{-2.17}. \tag{22}$$

In experiments with ground-based arrays the reconstructed number of muons and electrons are plotted in a $\ln N_\mu - \ln N_e$ plane to recover the energy and mass of the primary particle. This diagram, when combined with detailed shower simulations, proved to be a powerful tool for extracting information on primary mass.

Therefore, it is interesting to study the electron-to-muon ratio at shower maximum

$$\frac{N_e}{N_\mu} \approx 35.1 \cdot \left(\frac{E_0}{A}\right)^{0.15}. \tag{23}$$

with the energy in PeV. This ratio depends on the energy per nucleon E_0/A of the primary particle, thus showing that N_e/N_μ can be used to infer the mass of the primary particle if the energy is measured with a different, independent observable.

We can use the relation (23) to investigate the sensitivity of EAS arrays to the primary mass A [38,46]

$$\lg\left(\frac{N_e}{N_\mu}\right) = 1.54 + 0.15 \cdot \lg\left(\frac{E_0}{1\ \text{PeV}}\right) - 0.065 \cdot \ln A = C - 0.065 \cdot \ln A \tag{24}$$

If the energy is reconstructed from another independent observable, the mass of the primary CR can be determined by measuring the ratio N_e/N_μ. Therefore, the relative error on the electron-to-muon ratio is

$$\frac{\Delta(N_e/N_\mu)}{N_e/N_\mu} \sim 0.15\left[\frac{\Delta E_0}{E_0} + \frac{\Delta A}{A}\right] \sim 0.15\left[\frac{\Delta A}{A}\right] \tag{25}$$

with the consequence that to measure the elemental composition with a resolution of one unit in $\ln A$ the relative error on N_e/N_μ must be $\approx 15\%$. A resolution of one unit in $\ln A$ in principle allows to reconstruct 4 (or 5 ?) different mass groups: p, He, CNO, MgSi (?) and Fe. The large shower-to-shower fluctuations often only allow one to trace the light and heavy components or the parameter $\langle \ln A \rangle$ with energy. Similarly, from the relation (17) follows that the position of the shower maximum must be determined with a resolution of about one radiation length $X_0 \sim 37$ g/cm^2 to have a resolution in $\ln A$ of one unit.

4. Reconstruction of the Energy and Mass of the Primary Particle

The crucial point in air shower observations with EAS arrays is the reconstruction of the primary particle properties (especially energy and mass number) from the measured quantities. In fact, analysis of shower data consists in the disentanglement of a threefold problem involving primary energy, primary mass and modelling of hadronic interactions (for a discussion about hadronic interactions in CR physics see, as an example, refs. [47,48] and references therein). An intrinsic ambiguity affects the interpretation of data. Different combinations of the two following elements can produce similar showers. As an example, a *"short"* shower can be produced by a large cross-section, high inelasticity or heavy primary mass. On the contrary, a *"long"* shower, penetrating deeper in the atmosphere, can be produced by small cross-section, low inelasticity or light mass.

(1) shower development, mainly governed by the inelasticity and by the inelastic cross section
(2) elemental composition of the primary flux, that we don't know and want to measure

Strictly speaking, when operating with shower arrays there are no observables directly related to the mass of the primary particle, and its measurement is very indirect. We note, however, that the Cherenkov light emitted by a primary heavy nucleus high up in the atmosphere (the so-called *"direct Cherenkov light"*) is directly related to the charge (and therefore to the mass) of the primary particle [49,50]. Since this light is proportional to Z^2, heavy nuclei are more suited for detection. Charge resolution is about 10% for $Z > 10$. The main limitation is that it can only be used over a small energy range for each atomic charge Z.

The majority of experiments with shower arrays can therefore apply only 'statistical' methods according to a classical scheme:

1. From the experimental data, via some phenomenological functions determined by Monte Carlo simulations for the particular array, the measured observables (N_e, N_{hadr}, N_μ, X_{max}, ...) are reconstructed.
2. The distributions of such quantities are compared with those extracted from a detailed simulation of the EAS development in the atmosphere in which a trial CR spectrum is used.
3. The input spectrum is varied in order to optimize the agreement between the reconstructed and calculated distributions of measured observables.

Therefore, a typical data analysis consists in finding a combination of primary energy spectrum, elemental composition and hadronic interaction characteristics to obtain a consistent description of the experimental results. Clearly, this is not a measurement, but only a consistency check of some trial models. In case of discrepancy, it is difficult to identify the origin; in case of agreement, is the parameter combination unique?

Due to the reduced resolution in the measurement of the primary mass (see Section 3), the majority of shower arrays displayed the results only as a function of the total energy per particle with the so-called *"all-particle"* energy spectrum, that is, as a function of the total energy per nucleus, and not per nucleon. Any tentative to infere informations about elemental composition are limited, at most, to study the evolution of the *"light"* ("proton-like") or *"heavy"* ("iron-like") components as a function of the energy, with results which critically depend on Monte Carlo predictions.

In the last two decades, a number of multi-component experiments have started to measure, with high statistics, at the same time, different shower observables, on an event-by-event basis. This fact allowed to exploit sophisticated analysis techniques to infer the characteristics of the primary particle by measuring the correlation between different components (for a review see, for example, the Refs. [39,51] and references therein).

In a nutshell,

- *How to obtain the energy spectrum in shower arrays?*
  This is the first step in the analysis of CR data. We measure the spectrum in one observable and make a conversion to the energy spectrum. The observable typically used is the shower size because, as discussed in Section 3.2, the number of electrons at shower maximum is nearly independent on the primary mass: $N^A_{e|max} \approx N^p_{e|max}$. However, surface detectors are usually located deep in the atmosphere and do not measure the number of electrons at shower maximum. Beyond the maximum, the number of electrons is a mass-sensitive parameter, with a larger electron number for air showers initiated by light primaries, according to a relation of the type

$$N_e(E, A) = \alpha(A) \cdot E^{\beta(A)} \tag{26}$$

where the parameters α and β depend on the primary mass A. This implies a degeneracy in the reconstruction procedure because to recover the primary energy from the size spectrum we must assume a given elemental composition to be measured. If the composition changes in the investigated energy range, the relationship between the measured electron size and inferred energy will also vary. The number of electrons in the core region has been used in some experiment, as well as the particle density at a suitable given distance from the shower axis, in some large arrays (see, for example, the Refs. [11,52]). In both cases this densities, according to Monte Carlo simulations, are nearly independent of the primary mass.

- *How do we measure elemental composition at ground?*

The inelastic cross-section σ_{inel}^{Fe-Air} of iron at 1 PeV is about six times larger than for protons of equal energy. Hence, nuclei develop showers higher in atmosphere (smaller X_{max}) than protons, dissipating their energy much faster. Due to the shorter interaction length and the smaller energy per nucleon and because of the reduced attenuation of the muon component, nuclei-induced showers contain less particles in the e.m. component deep in the atmosphere, but they carry more muons than a proton shower of the same energy. This is the basis of the electron-muon correlation method. Therefore, as discussed in Section 3.4, the measurement of electron and muon contents simultaneously (with their fluctuations) has become the first and most commonly employed technique to infer the CR elemental composition with arrays. However, intrinsic shower to shower fluctuations limit mass resolution to a few mass groups (see Section 3.4) and electron and muon numbers are not independent. In addition, the muon component is heavy dependent on the details of the hadronic interactions and the results strongly depend on the particular model used to interpret the data.

The other common technique, below 10^{18} eV, involves the observation of the Cherenkov light and the study of its shape. In fact, the characteristics of the photon distribution depend on the depth of the shower maximum, therefore on the mass of the primary particle. The overall Cherenkov intensity provides a calorimetric measurement of the CR energy. Cherenkov light has been measured, for instance, in hybrid experiments by ARGO-YBJ and Tunka apparatus.

The KASCADE multi-component array was the first experiment that claimed the measurement of the energy spectra of 5 different mass groups (p, He, CNO, MgSi, Fe) through a complex unfolding of the N_e/N_μ diagram [22–24]. In the last two decades, other multi-component experiments measured a number of observables that, in principle, are mass-sensitive: steepness of the lateral distribution, characteristics of shower core region, distribution of the relative arrival times and angles of incidence of the muon component, characteristics of the lateral distribution of high energy muons (the so-called "*muon bundles*") measured underground, pulse shape and lateral distribution of the air Cherenkov light, depth X_{max} of the shower maximum (see, for example, ref. [15,17]). But the study of the muon component has remained the most used technique.

5. Elemental Composition in the 10^{14} to 10^{18} eV Region

Several experimental results associate the knee with the bending of the light component (p and He), and are compatible with a rigidity-dependent cut-off [22–24,53–55]. However, the flux of the different components vary significantly depending on the interaction model used to interpret the data [22–24]. On the contrary, other results (in particular those obtained by arrays located at high altitudes) seem to indicate that the knee of the all-particle energy spectrum is due to heavier nuclei and that the light component cuts off well below 1 PeV [21,26,31,52,53,56].

In this section the measurements of the light component energy spectrum in the 10^{14} to 10^{18} eV region will be presented by using different plots to point out the conflicting results between the experiments.

In Figure 5 the energy spectra of the light component as measured by Tibet ASγ [25,31] and ARGO-YBJ [52] are shown. Both experiments are located in the YangBaJing Cosmic Ray Laboratory in Tibet (China) at 4300 m a.s.l. and did not exploit a measurement of the muon component to determine the elemental composition of the primary CR flux.

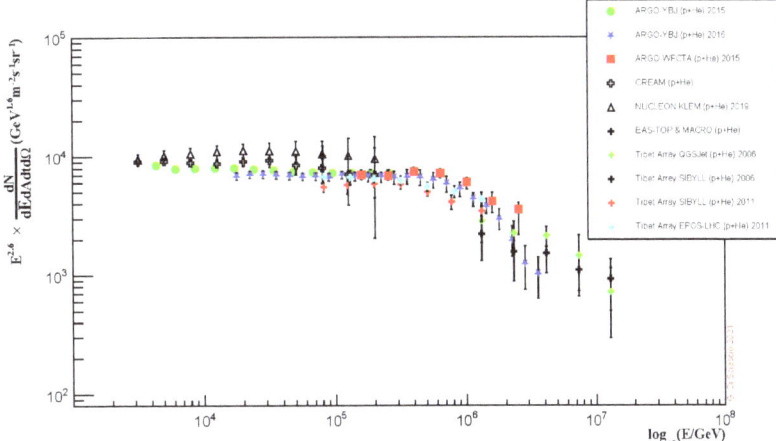

Figure 5. Energy spectra of the light (p+He) component as measured by Tibet ASγ [25,31] and ARGO–YBJ [52] experiments with different techniques and analyses, compared with results obtained in direct observations by CREAM [57] and NUCLEON [58].

The Tibet ASγ Collaboration reconstructed the energy spectrum studying the shower core region with a burst detector as well as with emulsion chambers. The ARGO-YBJ experiment measured the CR energy spectra exploiting completely different and independent approaches [52]:

- *'Digital-Bayes' analysis*, based on the strip multiplicity, that is, the picture of the EAS provided by the RPC strip/pad system, in the few TeV–300 TeV energy range. The selection of light elements is based on the characteristics of the charged particle lateral distribution [30,59].
- *'Analog-Bayes' analysis*, based on the RPC charge readout [60], covers the 30 TeV–10 PeV range. The energy is reconstructed (as in the previous analysis), on a statistical basis, by using a bayesian approach.
- *'Hybrid measurement'*, carried out by ARGO-YBJ and a wide field of view Cherenkov telescope, a prototype of the LHAASO telescopes, in the 100 TeV–3 PeV region. The selection of (p+He)-originated showers is based on two observables, the shape of the Cherenkov image and the particle density in the core region measured by the ARGO-YBJ central carpet. The energy is reconstructed by the telescope with a resolution better than 20% [56,61].

All the results are in excellent agreement. In the ARGO-YBJ experiment the selection of (p+He)-originated showers is performed not by means of an unfolding procedure after the measurement of electronic and muonic sizes, but on an event-by-event basis exploiting showers topology, that is, the lateral distribution of charged secondary particles. This approach is made possible by the full coverage of the central carpet, the high segmentation of the read-out and the high altitude location of the experiment that retains the characteristics of showers lateral distribution in the core region. The contamination of nuclei heavier than helium is estimated smaller than 15% at 1 PeV in all analyses.

In Figure 5 the direct measurements reported by CREAM [57] and NUCLEON [58] are also shown. ARGO-YBJ is the only experiment that traced the (p+He) spectrum across the knee starting from an energy so low (≈TeV) to overlap with direct measurements and to cross-calibrate the fluxes on a wide energy range (5–250 TeV). These results show that, when indirect measurements are capable of selecting almost pure beams, their findings are in fair agreement with direct ones and confirm that current simulation models provide a satisfactory description of the EAS development in the atmosphere. The cross-calibration of fluxes in this energy range, where the boundary line between 'direct' and 'indirect' measurements is uncertain, is very important. The low energy threshold allowed also a calibration of the absolute energy scale at a level of 10% exploiting the *Moon Shadow* technique in the 1–30 TeV/Z range [20].

As can be seen from the figure, the observations of Tibet ASγ and ARGO-YBJ are in good agreement each other showing that the knee of the (p+He) energy spectrum is at $\approx$500–700 TeV, well below the energy of knee in the all-particle spectrum. Similar conclusions have been obtained by the BASJE-MAX experiment located at 5200 m asl [21] and by EAS-TOP at 2000 m asl [53] and by CASA-MIA at 1450 m asl [26].

In Figure 6 the energy spectra of the light component reconstructed by the KASCADE experiment [22–24] with two different hadronic interaction models are added for comparison. The energy threshold is about 1 PeV and the experiment was located at sea level. KASCADE did use of a complex unfolding procedure to recover the elemental composition from the N_e–N_μ diagram in terms of 5 mass groups (p, He, CNO, MgSi, Fe). As can be seen from the figure, both the spectra are at variance with the results obtained by Tibet ASγ and ARGO-YBJ, suggesting that the knee of the CR all-particle spectrum at a few PeV is due to the bending of the light component.

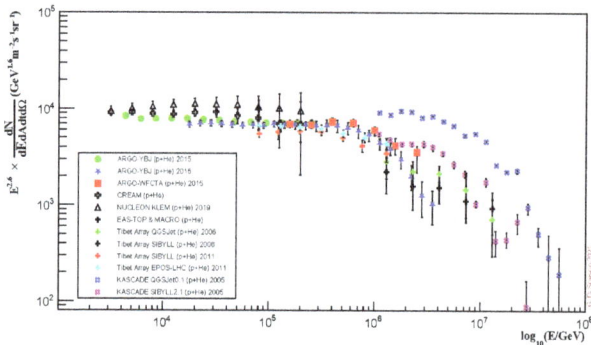

Figure 6. The energy spectra shown in Figure 5 compared with the results obtained by the KASCADE experiment by using two different interaction models to interpret data [22–24].

All measurements of the light component up to about 10^{18} eV are summarized in the Figure 7 and compared with the parametrization provided by Horandel [62]. Roughly speaking, we can separate the experiments that measured the (p+He) energy spectrum in the PeV range in 2 different groups

1. arrays located at extreme altitude (BASJE-MAS at 5200 m asl, ARGO-YBJ and Tibet ASγ at 4300 m asl) observing a composition at the knee heavier than (p+He). These experiments did not exploit the measurement of the muon component to recover the elemental composition;
2. arrays located deeper in the atmosphere (KASCADE, KASCADE-Grande and Ice-Top/Icecube) reporting evidence that the light cut-off is located at a few PeV. In this case both low and high energy muons have been used in the classical study of N_e–N_μ correlation.

Measurements exploiting the longitudinal profile of the showers with Cherenkov detectors are conflicting too. The results obtained with the ARGO-YBJ hybrid detector (carpet and Cherenkov telescope) are in agreement with those of the carpet only, whereas the observations of Cherenkov light by Tunka-133 are consistent with KASCADE and KASCADE-Grande findings.

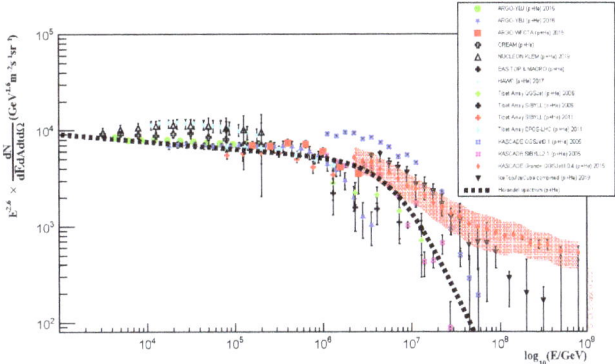

Figure 7. The energy spectra shown in Figure 6 compared with the results obtained by HAWC in the 10–100 TeV region [63] and by KASCADE–Grande [27] and IceTop/IceCube [29] combined above the PeV. The parametrization of the light component provided by Hörandel [62] is also shown.

In Figure 8 the all-particle energy spectrum measured by several experiments in the energy region 10^{16}–10^{18} eV is shown. As can be seen, the spectrum cannot be fitted by a single power law. We observe a spectral hardening at $\sim 2 \times 10^{16}$ eV and a steepening at $\sim 10^{17}$ eV. This result was first pointed out by KASCADE-Grande experiment [64,65], then more firmly assessed with higher statistics and precision by Tunka-133 and IceTop-73 [28,29], in particular for the feature at $\sim 2 \times 10^{16}$ eV.

The *light* (p+He) and *heavy* (C-Fe group) components measured by KASCADE-Grande are also shown. A knee is observed in the heavy component of CRs at $E = 10^{16.92 \pm 0.04}$ eV, which coincides within the uncertainties with the change of the slope in the all-particle energy spectrum around 10^{17} eV. The spectral index changes from -2.76 ± 0.02 below the knee to -3.24 ± 0.05 above. At slightly higher energies ($E = 10^{17.08 \pm 0.09}$ eV), the light component shows a hardening of the slope, with the spectral index changing from -3.25 ± 0.06 below this ankle-like structure to -2.79 ± 0.09 above. The positions of the changes of the slope as well as the intensities of the different components depend on the interaction model adopted to interpret the data. The knee in the heavy component seems visible also in the all-particle spectrum, as it is the dominant component.

The results obtained by Tunka-133, measuring the Cherenkov light deep in the atmosphere, suggest that the mass composition becomes heavier in the energy range 10^{16}–3×10^{16} eV, then stays heavy till 10^{17} eV, where the composition starts becoming lighter. IceTop/IceCube, exploiting the high-energy muons underground, indicates an increase of $\langle \ln A \rangle$ in the energy range 10^{16}–3×10^{17} eV [65]. However, as discussed in the previous sections, the average logarithmic mass of CR $\langle \ln A \rangle$ is used to describe the evolution of the composition as a function of energy when the mass resolution of the experiments do not allow a discrimination between different mass groups.

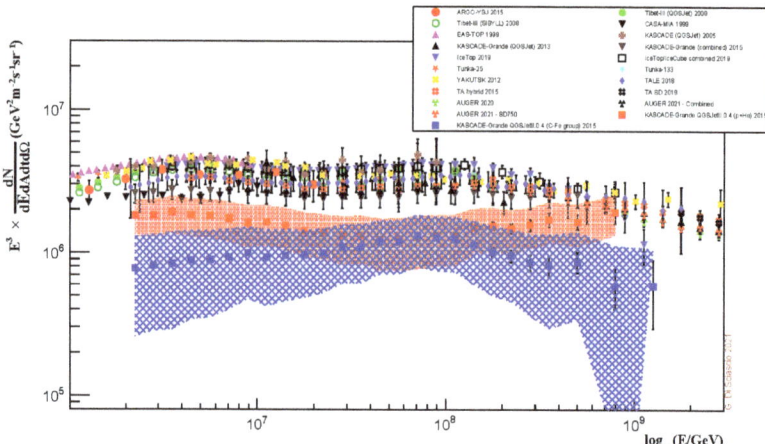

Figure 8. The all–particle energy range in the 'transition' region measured by different experiments. The *light* and *heavy* components measured by KASCADE–Grande are also shown.

In Figure 8 recent measurements of the all-particle energy spectrum down to about 100 PeV by the Pierre Auger Observatory are also reported. These observations suggest that the second knee is not a sharp feature, but a softening that extends in the interval 100–200 PeV [66]. Preliminary results based on the distribution of the depth of the shower maximum are consistent with a spectrum dominated by heavy nuclei in the 10^{17} eV range becoming lighter with increasing energy [67].

The findings of Tunka-133, IceTop/IceCube and Auger are qualitatively in agreement with KASCADE-Grande. Despite the large uncertainty in the absolute composition, a common general trend is reported, composition gets heavier through the knee region and becomes lighter approaching the ankle.

In conclusion, the observations of the different ground-based arrays show two conflicting results regarding the maximum energy at which the light component is accelerated in CR sources. This disagreement is summarized in Figure 9 where the ARGO-YBJ results are compared with the KASCADE-Grande light and heavy spectra. These results cannot be reconciled and show the existence of a still unknown systematic uncertainty that, as discussed in previous sections, could be due to the different array characteristics (altitude, coverage), the observables used (muons or shower core characteristics), and the dependence on hadronic interaction models. These are certainly among the major sources of systematic errors that affect the interpretation of shower array measurements (for a recent discussion see, for instance, refs. [68,69]), although recent re-analyses of the KASCADE-Grande data with the latest versions of the post-LHC codes confirm previous results [70].

Important information could be deduced, in principle, by the measurement of the flux of atmospheric neutrinos, sensitive to the spectrum of parent CRs. In particular, the tail of the spectrum of atmospheric neutrinos is mainly shaped by the parent protons rather than by heavier element. As a consequence, we expect different predictions for the flux of atmospheric neutrinos according to ARGO-YBJ and KASCADE proton energy spectra, predictions that, in principle, can be checked at energies $E_\nu \geq 100$ TeV if the atmospheric neutrinos could be properly identified. Unfortunately, in this energy region the total neutrino flux detected by IceCube departs from the existing predictions for atmospheric neutrinos suggesting the onset of an astrophysical component. The origin of such neutrinos is still unknown and current experimental uncertainties do not allow to draw clear conclusions [71].

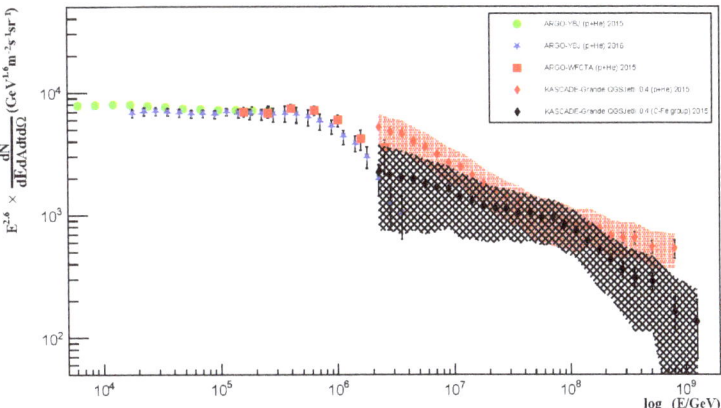

Figure 9. The energy spectra of the light component measured by ARGO–YBJ compared to the *light* and *heavy* components measured by KASCADE–Grande.

6. What's Next

The experimental situation in the 100 TeV–100 PeV energy region must be clarified to solve the longstanding problem of the origin of the knee and to give solid foundations to CR models up to the highest observed energies. A new experiment, able to measure, at the right altitude and with high statistics, the elemental composition exploiting the techniques used so far in different apparatus, is mandatory to investigate the unknown uncertainties affecting the results so far obtained by shower arrays.

The only experiment that meets these requirement is LHAASO, a new multi-component array developed starting from the experience of the high altitude experiment ARGO-YBJ. The apparatus is located at high altitude (4410 m asl, 600 g/cm^2) in the Daochen site, Sichuan province, P.R. China. LHAASO is expected to measure the energy spectrum, the elemental composition and the anisotropy of CRs in the energy range between 10^{12} and 10^{17} eV [11,12,72,73]. The experiment is constituted by a 1 km^2 dense array of plastic scintillators and muon detectors. At the center of the array a 300×300 m^2 water Cherenkov facility will allow the detection of TeV showers. An array of 18 wide field of view Cherenkov telescopes will image the longitudinal profile of events. Neutron monitors will study the hadronic component in the core of air showers. LHAASO will study CR physics with different detectors and techniques starting from the TeV range, thus overlapping direct measurements in a wide interval. In Tables 1 and 2 the characteristics of the LHAASO-KM2A array are compared with other experiments. As can be seen, LHAASO will operate with a coverage of ∼0.5% over a 1 km^2 area. The sensitive area of muon detectors is unprecedented (more than 40,000 m^2), about 17 times larger than the CASA-MIA experiment, with a coverage of about 5% over 1 km^2. For the first time the N_e/N_μ correlation will be studied at high altitude with high statistics. This suite of independent instruments will also allow a deep investigation of the characteristics of the hadronic interaction models. The capability of hybrid measurements with Cherenkov telescopes operated in combination with a shower array have been demonstrated by the ARGO-YBJ measurement of the light component energy spectrum.

In addition, LHAASO will act simultaneously as a wide aperture (∼2 sr), continuously-operated gamma-ray telescope in the energy range between 10^{11} and 10^{15} eV. The first results obtained during the first year of data taking with only a portion of the apparatus opened for the first time the PeV sky to observations, showing that the Northern hemisphere contains a lot of galactic PeVatrons.

Other projects under way to investigate, with a much higher energy threshold, the high energy tail of the galactic spectrum and the transition region are HiSCORE [74] and GRAND [75].

7. Conclusions

The results obtained by different experiments in the 10^{14} to 10^{18} eV region can be summarized as follows:

- Knee energy region

    1. All experiments observe an all-particle knee at $\approx 4 \times 10^{15}$ eV.
    2. The absolute fluxes are in good agreement with each other and with the direct measurements.
    3. The elemental composition is conflicting. Experiments located at high and extreme altitude (BASJE-MAS, Tibet ASγ, ARGO-YBJ, EAS-TOP and CASA-MIA) reported evidence that the knee of the (p+He) component is below 1 PeV and that the composition at the all-particle knee energy is dominated by heavier nuclei. Experiments located deeper in atmosphere (KASCADE, KASCADE-Grande, IceTop/IceCube, Tunka-133) reported evidence that the proton knee is at the same energy of the all-particle knee.
    4. A 10^{-3}–10^{-4} Large Scale Anisotropy (LSA) amplitude is found at TeV energies [76].
    5. A 10^{-4} Medium Scale Anisotropy (MSA) amplitude is observed at TeV energies [76].

- Transition region 10^{16}–10^{18} eV

    1. The all-particle energy spectrum measured by different experiments are in good agreement within the systematics and with the measurements of UHE experiments.
    2. A concave region is observed above 2×10^{16} eV with a steepening at $\sim 10^{17}$ eV.
    3. The dipole component of the LSA is smaller than 10^{-2}.

The observed features in the all-particle energy spectrum seem to be consistent with the bending of different components in a rigidity-based scenario. However, rigidity models can be

- *rigidity-acceleration* models, that is, the knee can be an acceleration feature, a source property, related to the maximum energy of particle acceleration inside the CR sources;
- *rigidity-confinement* models, that is, the knee is related to inefficient confinement of particles in the galaxy. In this case, the galaxy could contain 'super-PeVatrons', astrophysical objects able to accelerate particles well beyond the PeV.

The first PeVatrons observed in the northern hemisphere by the LHAASO experiment show that SNRs are probably not the main sources of PeV CRs in our galaxy. The observation of sources emitting photons above the PeV in the North suggests the need of a wide field of view instrument in the Southern Hemisphere to monitor the Inner Galaxy and the Galactic Center looking for super-PeVatrons (SWGO [77], STACEX [78]) to operate with CTA-South [79].

In the coming years, the LHAASO experiment is expected to be able to measure the energy spectra of different mass groups up to 10^{17} eV and to determine the energy of the proton knee, thus clarifying the origin of the knee in the all-particle spectrum. The suite of independent instruments that will be operated will also allow a deep study of the characteristics of the hadronic interaction models and to investigate the uncertainties related to the main techniques used to recover the elemental composition.

Funding: This research received no external funding.

Institutional Review Board Statement: Not applicable.

Informed Consent Statement: Not applicable.

Data Availability Statement: Not applicable.

Conflicts of Interest: The authors declare no conflict of interest

References

1. Battistoni, G.; Grillo, A.F. Introduction to High-Energy Cosmic Ray Physics. In Proceedings of the ICTP School on Nonaccelerator Particle Astrophysics, Trieste, Italy, 17–28 July 1995; Preprint INFN/AE—96/05; pp. 341–374.
2. Di Sciascio, G. Detection of Cosmic Rays from ground: An Introduction. *J. Phys. Conf. Ser.* **2019**, *1263*, 012002. [CrossRef]
3. Stanev, T.C.R. Cosmogenic neutrinos and gamma rays. *Physique* **2014**, *15*, 349–356. [CrossRef]
4. Williams, R.W. The structure of the large cosmic-ray air showers. *Phys. Rev.* **1948**, *74*, 1689–1706. [CrossRef]
5. Linsley, J. Spectra, anisotropies and composition of cosmic rays above 1000 GeV. *Proc. ICRC* **1983**, *12*, 135L.
6. Kulikov, G.V.; Khristiansen, G.B. On the size spectrum of extensive air showers. *Sov. Phys. JETP* **1959**, *35*, 441–444.
7. Miura, I.; Hasegawa, H. Spectra of the Size and the Total Number of Mu-Mesons in EAS. *J. Phys. Soc. Jpn.* **1962**, *17*, 84.
8. Peters, B. Primary cosmic radiation and extensive air showers. *Il Nuovo Cimento* **1961**, *22*, 800. [CrossRef]
9. Drury, L.O. Origin of cosmic rays. *Astropart. Phys.* **2012**, *39–40*, 52–60. [CrossRef]
10. Gabici, S.; Gaggero, D.; Zandanel, F. Can supernova remnants accelerate protons up to PeV energies? *arXiv* **2016**, arXiv:1610.07638.
11. Cao, Z.; LHAASO Collaboration. Ultrahigh-energy photons up to 1.4 petaelectronvolts from 12 γ-ray Galactic sources. *Nature* **2021**, *594*, 33–36. [CrossRef]
12. Cao, Z.; LHAASO Collaboration. Peta–electron volt gamma-ray emission from the Crab Nebula. *Science* **2021**, *373*, 425–430.
13. Bartoli, B.; Bernardini, P.; Bi, X.J.; Branchini, P.; Budano, A.; Camarri, P.; Cao, Z.; Cardarelli, R.; Catalanotti, S.; Chen, S.Z.; et al. Identification of the TeV gamma-ray source ARGO J2031+4157 with the Cygnus Cocoon. *ApJ* **2014**, *790*, 152. [CrossRef]
14. Aharonian, F.; Yang, R.; de Ona Wilhelmi, E. Massive stars as major factories of Galactic cosmic rays. *Nat. Astron.* **2019**, *3*, 561–567. [CrossRef]
15. Spurio, M. *Probes of Multimessenger Astrophysics*; Springer International Publishing: Cham, Switzerland, 2018.
16. Gaisser, T.K.; Engel, R.; Resconi, E. *Cosmic Rays and Particle Physics*; Cambridge University Press: Cambridge, UK, 2016.
17. Grieder, P.K.F. *Extensive Air Showers*; Springer International Publishing: Bern, Switzerland, 2010.
18. Longair, M.S. *High Energy Astrophysics*; Cambridge University Press: Cambridge, UK, 1981.
19. Aloisio, R. *Multiple Messengers and Challenges in Astroparticle Physics*; Springer International Publishing: Cham, Switzerland, 2018.
20. Bartoli, B.; Bernardini, P.; Bi, X.J.; Cao, Z.; Catalanotti, S.; Chen, S.Z.; Chen, T.L.; Cui, S.W.; Dai, B.Z.; D'Amone, A.; et al. Observation of the cosmic ray moon shadowing effect with the ARGO-YBJ experiment. *Phys. Rev. D* **2011**, *84*, 022003. [CrossRef]
21. Ogio, S.; Kakimoto, F.; Kurashina, Y.; Burgoa, O.; Harada, D.; Tokuno, H.; Yoshii, H.; Morizawa, A.; Gotoh, E.; Nakatani, H.; et al. The energy spectrum and the chemical composition of primary cosmic rays with energies from 10^{14} to 10^{16} eV. *ApJ* **2004**, *612*, 268. [CrossRef]
22. Antoni, T.; Apel, W.D.; Badea, A.F.; Bekk, K.; Bercuci A.; Blümer, J.; Bozdog, H.; Brancus, I.M.; Chilingarian, A.; Daumiller, K.; et al. KASCADE measurements of energy spectra for elemental groups of cosmic rays: Results and open problems. *Astropart. Phys.* **2005**, *24*, 1. [CrossRef]
23. Apel, W.D.; Arteaga, J.C.; Badea, A.F.; Bekk, K.; Blümer J.;Bozdog, H.; Brancus, I.M.; Brüggemann, M.; Buchholz, P.; Cossavella, F.; et al. KASCADE Collaboration. Energy spectra of elemental groups of cosmic rays: Update on the KASCADE unfolding analysis. *Astropart. Phys.* **2009**, *31*, 86. [CrossRef]
24. Apel, W.D.; Arteaga-Velázquez, J.C.; Bekk, K.; Bertaina, M.; Blümer, J.; Bozdog, H.; Brancus, I.M.; Cantoni, E.; Chiavassa, A.; Cossavella, F.; et al. KASCADE-Grande measurements of energy spectra for elemental groups of cosmic rays. *Astropart. Phys.* **2013**, *47*, 54. [CrossRef]
25. Amenomori, M.; Bi, X.J.; Chen, D.; Cui, S.W.; Danzengluobu; Ding, L.K.; Ding, X.H.; Fan, C.; Feng, C.F.; Fenget, Z.; et al. Cosmic-ray energy spectrum around the knee obtained by the Tibet experiment and future prospects. *Adv. Space Res.* **2011**, *47*, 629. [CrossRef]
26. Glasmacher, M.A.K.; Catanese, M.A.; Chantell M.C.; Covault, C.E.; Cronin, J.W.; Fick, B.E.; Fortson, L.F.; Fowler, J.W.; Green, K.D.; Kieda, D.B.; et al. CASA-MIA Collaboration. The cosmic ray composition between 10^{14} and 10^{16} eV. *Astropart. Phys.* **1999**, *12*, 1. [CrossRef]
27. Apel, W.D.; Arteaga-Velázquez, J.C.; Bekk, K.; Bertainaet, M.; Bluemer, J.; Bozdog, H.; Brancus, I.M.; Buchholz, P.; Cantoni, E.; Chiavassa, A.; et al. The spectrum of high-energy cosmic rays measured with KASCADE-Grande. *Astropart. Phys.* **2012**, *36*, 183. [CrossRef]
28. Prosin, V.V.; Berezhnev, S.F.; Budnev, N.M.; Brückner, M.; Chiavassa, A.; Chvalaev, O.A.; Dyachok, A.V.; Epimakhov, S.N.; Gafarov, A.V.; Gress, O.A.; et al. TUNKA Collaboration. Results from Tunka-133 (5 years observation) and from the Tunka-HiSCORE prototype. In Proceedings of the 5th Roma International Conference on Astro-Particle physics (RICAP 14), Sicily, Italy, 30 September–3 October 2014.
29. Aartsen, M.G.; Ackermann, M.; Adams, J.; Aguilar J.A.; Ahlers, M.; Ahrens, M.; Alispach, C.; Andeen, K.; Anderson, T.; Ansseau, I.; et al. ICETOP Collaboration. Cosmic ray spectrum and composition from PeV to EeV using 3 years of data from IceTop and IceCube. *Phys. Rev. D* **2019**, *100*, 082002. [CrossRef]

30. Bartoli, B.; Bernardini, P.; Bi, X.J.; Bleve, C.; Bolognino I.; Branchini, P.; Budano, A.; Calabrese Melcarne, A.K.; Camarri, P.; Cao, Z.; et al. Light-component spectrum of the primary cosmic rays in the multi-TeV region measured by the ARGO-YBJ experiment. *Phys. Rev. D* **2012**, *85*, 092005. [CrossRef]
31. Amenomori, M.; Ayabe, S.; Chen, D.; Cui, S.W.; Danzengluobu; Ding, L.K.; Ding, X.H.; Feng, C.F.; Feng, Z.Y.; Gao, X.Y.; et al. Are protons still dominant at the knee of the cosmic-ray energy spectrum? *Phys. Lett. B* **2006**, *632*, 58–64. [CrossRef]
32. Heitler, W. *The Quantum Theory of Radiation*; Clarendon Press: Oxford, UK, 1944.
33. Matthews, J. A Heitler model of extensive air showers. *Astropart. Phys.* **2005**, *22*, 387. [CrossRef]
34. Letessier-Selvon, A.; Stanev, T. Ultrahigh energy cosmic rays. *Rev. Mod. Phys.* **2011**, *83*, 907–942. [CrossRef]
35. Heck, D.; Knapp, J.; Capdevielle, J.N.; Schatz, G.; Thouw, T. *CORSIKA: A Monte Carlo Code to Simulate Extensive Air Showers*; Forschungszentrum Karlsruhe GmbH: Karlsruhe, Germany, 1998.
36. Allison, J.; Amako, K.; Apostolakis, J.; Arce, P.; Asai, M.; Aso, T.; Bagli, E.; Bagulya, A.; Banerjee, S.; Barrand, G.; et al. Recent developments in Geant4. *NIM* **2016**, *A835*, 186–225. [CrossRef]
37. Sciutto, S.J. AIRES: A system for air shower simulations. *arXiv* **2019**, arXiv:astro-ph/9911331.
38. Horandel, J.R. Cosmic Rays from the Knee to the Second Knee: 10^{14} to 10^{18} eV. *Mod. Phys. Lett. A* **2007**, *22*, 1533–1551. [CrossRef]
39. Kampert, K.H.; Unger, M. Measurements of the cosmic ray composition with air shower experiments. *Astropart. Phys.* **2012**, *35*, 660–678. [CrossRef]
40. Mollerach, S.; Roulet, E. Progress in high-energy cosmic ray physics. *Progr. Part. Nucl. Phys.* **2018**, *98*, 85–118. [CrossRef]
41. Linsley, J. Structure of large air showers at depth 834 g/cm^{-2}: Applications. In Proceedings of the 15th International Cosmic Ray Conference, Plovdiv, Bulgaria, 13–26 August 1977; Volume 12, p. 89.
42. Fukui, S.; Hasegawa, H.; Matano, T.; Miura, I.; Oda, M.; Suga, K.; Tanahashi, G.; Tanaka, Y. A study on the structure of the extensive air shower. *Suppl. Prog. Theor. Phys.* **1960**, *16*, 1–53. [CrossRef]
43. Matano, T.; Miura, I.; Nagano, M.; Oda, M.; Shibata, S.; Tanaka, Y.; Tanahashi, G.; Hasegawa, H. Extensive air showers—Studies of Tokyo group. In Proceedings of the 8th International Cosmic Ray Conference, Jaipur, India, 2–14 December 1963; Volume 4, p. 129.
44. Vernov, S.N.; Khristiansen, G.B.; Abrosimov, A.M.; Atrashkevich, V.B.; Beliaeva, M.G. A descriptior of a modified complex installation for investigation of extensive air showers and new experimental data obtained by means of this installation. In Proceedings of the 8th International Cosmic Ray Conference, Jaipur, India, 2–14 December 1963; Volume 4, p. 173
45. Linsley, J.; Scarsi, L.; Rossi, B. Energy spectrum and structure of large air showers. *J. Phys. Soc. Japan* **1962**, *17*, 91
46. Horandel, J.R. Cosmic-ray composition and its relation to shock acceleration by supernova remnants. *Adv. Space Res.* **2008**, *41*, 442–463. [CrossRef]
47. Lipari, P. Cosmic rays and hadronic interactions. *C. R. Phys.* **2014**, *15*, 357–366. [CrossRef]
48. Riehn, F.; Engel, R.; Fedynitch, A.; Gaisser, T.K.; Stanev, T. Hadronic interaction model Sibyll 2.3d and extensive air showers. *Phys. Rev. D* **2020**, *102*, 063002. [CrossRef]
49. Kieda, D.B.; Swordy, S.P.; Wakely, S.P. A high resolution method for measuring cosmic ray composition beyond 10 TeV. *Astropart. Phys.* **2001**, *15*, 287–303. [CrossRef]
50. Aharonian, F.; Akhperjanian, A.G.; Bazer-Bachi, A.R.; Akhperjanian, A.G.; Angüner, E.O.; Backes, M.; Balenderan, S.; Balzer, A.; Barnacka, A.; Becherini, Y.; et al. HESS Collaboration. First ground-based measurement of atmospheric Cherenkov light from cosmic rays. *Phys. Rev. D* **2007**, *75*, 042004. [CrossRef]
51. Haungs, A.; Rebel, H.; Roth, M. Energy spectrum and mass composition of high-energy cosmic rays. *Rep. Prog. Phys.* **2003**, *66*, 1145. [CrossRef]
52. Di Sciascio, G. Main physics results of the ARGO-YBJ experiment. *Int. J. Mod. Phys. D* **2014**, *23*, 1430019. [CrossRef]
53. Aglietta, M.; Alessandro, B.; Antonioli, P.; Arneodo, F.; Bergamasco, L.; Bertaina, M.; Castagnoli, C.; Castellina, A.; Chiavassa, A.; Cini Castagnoli, G.; et al. EAS-TOP Collaboration. The cosmic ray primary composition in the "knee" region through the EAS electromagnetic and muon measurements at EAS-TOP. *Astropart. Phys.* **2004**, *21*, 583. [CrossRef]
54. Garyaka, A.P.; Martirosov, R.M.; Ter-Antonyan, S.V.; Nikolskaya, N.; Gallant, Y.A.; Jones, L.; Procureur, J. GAMMA Collaboration. Rigidity-dependent cosmic ray energy spectra in the knee region obtained with the GAMMA experiment. *Astropart. Phys.* **2007**, *28*, 169. [CrossRef]
55. Tanaka, H.; Dugad, S.R.; Gupta, S.K.; Jain A.; Mohanty, P.K.; Rao, B.S.; Ravindran, K.C.; Sivaprasad, K.; Tonwar, S.C.; Hayashi, Y.; et al. GRAPES Collaboration. Studies of the energy spectrum and composition of the primary cosmic rays at 100–1000 TeV from the GRAPES-3 experiment. *J. Phys. G Nucl. Part. Phys.* **2012**, *39*, 025201. [CrossRef]
56. Bartoli, B.; Bernardini, P.; Bi, X.J.; Cao, Z.; Catalanotti, S.; Camarri, P.; Cao, Z.; Cardarelli, R.; Catalanotti, S.; Chen, S.Z.; et al. Knee of the cosmic hydrogen and helium spectrum below 1 PeV measured by ARGO-YBJ and a Cherenkov telescope of LHAASO. *Phys. Rev. D* **2015**, *92*, 092005. [CrossRef]
57. Yoon, Y.S.; Ahn, H.S.; Allison, P.S.; Bagliesi, M.G.; Beatty, J.; Bigongiari, G.; Boyle, P.J.; Childers, J.T.; Conklin, N.B.; Coutu, S.; et al. CREAM Collaboration. Cosmic-ray proton and helium spectra from the first CREAM flight. *ApJ* **2011**, *728*, 122. [CrossRef]
58. Grebenyuk, V.; Karmanov, D.; Kovalev, I.; Kovalev, I.; Kudryashov, I.; Kurganov, A.; Panov, A.; Podorozhny, D.; Porokhovoy, S.; Sveshnikova, L.; et al. NUCLEON Collaboration. Energy spectra of abundant cosmic-ray nuclei in the NUCLEON experiment. *Adv. Space Res.* **2019**, *64*, 2546. [CrossRef]

59. Bartoli, B.; Bernardini, P.; Bi, X.J.; Cao, Z.; Catalanotti, S.; Chen, S.Z.; Chen, T.L.; Cui, S.W.; Dai, B.Z.; D'Amone, A.; et al. Cosmic ray proton plus helium energy spectrum measured by the ARGO-YBJ experiment in the energy range 3–300 TeV. *Phys. Rev. D* **2015**, *91*, 112017. [CrossRef]
60. Bartoli, B.; Bernardini, P.; Bi, X.J.; Branchini, P.; Budano, A.; Chen, S.Z.; Chen, T.L.; Cui, S.W.; Dai, B.Z.; D'Amone, A.; et al. The analog Resistive Plate Chamber detector of the ARGO-YBJ experiment. *Astropart. Phys.* **2015**, *67*, 47. [CrossRef]
61. Bartoli, B.; Bernardini, P.; Bi, X.J.; Bolognino, I.; Branchini, P.; Budano, A.; Calabrese Melcarne, A.K.; Camarri, P.; Cao, Z.; Cardarelli, R.; et al. Energy spectrum of cosmic protons and helium nuclei by a hybrid measurement at 4300 m asl. *Chinese Phys.* **2014**, *C38*, 045001. [CrossRef]
62. Hörandel, J.H. On the knee in the energy spectrum of cosmic rays. *Astropart. Phys.* **2003**, *19*, 193. [CrossRef]
63. Alfaro, R.; Alvarez, C.; Álvarez, J.D.; Arceo, R.; Avila Rojas, D.; Ayala Solares, H.A.; Barber, A.S.; Becerril, A.; Belmont-Moreno, E.; BenZvi, S.Y.; et al. HAWC Collaboration. All-particle cosmic ray energy spectrum measured by the HAWC experiment from 10 to 500 TeV. *Phys. Rev. D* **2017**, *96*, 122001. [CrossRef]
64. Bertaina, M.E.; Apel, W.D.; Hörandel, J.R.; Wommer, M.; Blumer, J.; Bozdog, H.; Brancus, I.M.; Buchholz, P.; Cantoni, E.; Chiavassa, A.; et al. KASCADE-Grande Collaboration, The cosmic ray energy spectrum in the range 10^{16}–10^{18} eV measured by KASCADE-Grande. *Astrophys. Space Sci. Trans.* **2011**, *7*, 229. [CrossRef]
65. Bertaina, M.E. Cosmic rays from the knee to the ankle. *C. R. Phys.* **2014**, *15*, 300–308. [CrossRef]
66. Abreu, P.; Aglietta, M.; Albury, J.M.; Almela, A.; Alvarez-Muñiz, J.; Alves Batista, R.; Anastasi, G.A.; Anchordoqui, L.; Andrada, B.; Allekotte, I.; et al. The energy spectrum of cosmic rays beyond the turn-down around 10^{17} eV as measured with the surface detector of the Pierre Auger Observatory. *Eur. Phys. J.* **2021**, *81*, 966. [CrossRef]
67. Bellido, J.; Aglietta, M.; Albury, J.M.; Allekotte, I.; Almeida Cheminant, K.; Almela, A.; Alvarez-Muñiz, J.; Alves Batista, R.; Anastasi, G.A.; Anchordoqui, L.; et al. (The Pierre Auger Collaboration). Depth of maximum of air-shower profiles at the Pierre Auger Observatory: Measurements above $10^{17.2}$ eV and Composition Implications. In Proceedings of the 35th International Cosmic Ray Conference (ICRC2017), Busan, Korea, 10–20 July 2017.
68. Cazon, L. Probing High-Energy Hadronic Interactions with Extensive Air Showers. In Proceedings of the 36th International Cosmic Ray Conference (ICRC2019), Madison, WI, USA, 24 July–1 August 2019; Volume 358, p. 5.
69. Cazon, L. Working Group Report on the Combined Analysis of Muon Density Measurements from Eight Air Shower Experiments. In Proceedings of the 36th International Cosmic Ray Conference (ICRC2019), Madison, WI, USA, 24 July–1 August 2019; Volume 358, p. 214.
70. Kang, D.; Haungs, A.; Apel, W.D.; Arteaga-Velázquez, J.C.; Beket, K.; Bertaina, M.; Blümer, J.; Bozdog, H.; Cantoni, E.; Chiavassa, A.; et al. Latest Results from the KASCADE-Grande Data Analysis. In Proceedings of the 36th International Cosmic Ray Conference (ICRC2019), Madison, WI, USA, 24 July–1 August 2019; Volume 358, p. 306.
71. Mascaretti, C.; Blasi, P.; Evoli, C. Atmospheric neutrinos and the knee of the cosmic ray spectrum. *Astropart. Phys.* **2020**, *114*, 22–29. [CrossRef]
72. Di Sciascio, G.; LHAASO Collaboration. The LHAASO experiment: From Gamma-Ray Astronomy to Cosmic Rays. *Nucl. Part. Phys. Proc.* **2016**, *279–281*, 166–173. [CrossRef]
73. Bai, X.; Bi, B.Y.; Bi, X.J.; Cao, Z.; Chen, S.Z.; Chen, Y.; Chiavassa, A.; Cui, X.H.; Dai, Z.G.; della Volpe, D.; et al. The Large High Altitude Air Shower Observatory (LHAASO) Science White Paper. *arXiv* **2019**, arXiv:1905.02773.
74. Tluczykont, M.; Hampf, D.; Horns, D.; Spitschan, D.; Kuzmichev, L.; Prosin, V.; Spiering, C.; Wischnewski, R.; HiSCORE Collaboration. The HiSCORE concept for gamma-ray and cosmic-ray astrophysics beyond 10 TeV. *Astropart. Phys.* **2014**, *56*, 42–53. [CrossRef]
75. Alvarez-Muniz, J.; Batista, R.A.; Bolmont, J.; Bolmont, J.; Bustamante, M.; Carvalho, W., Jr.; Charrier, D.; Cognard, I.; Decoene, V.; Denton, P.B.; et al. The giant radio array for neutrino detection (GRAND): Science and design. *Sci. China* **2020**, *63*, 219501. [CrossRef]
76. Di Sciascio, G.; Iuppa, R. *Homage to the Discovery of Cosmic Rays*; Perez-Peraza, J.A., Ed.; Nova Science Publishers: New York, NY, USA, 2013; Chapter 9, pp. 221–257.
77. Albert A.; Alfaro, R.; Ashkar, H.; Alvarez, C.; Álvarez, J.; Arteaga-Velázquez, J.C.; Ayala Solares, H.A.; Arceo, R.; Bellido, J.A.; BenZvi, S.; et al. Science Case for a Wide Field-of-View Very-High-Energy Gamma-Ray Observatory in the Southern Hemisphere. *arXiv* **2019**, arXiv:1902.08429v1.
78. Rodriguez-Fernandez, G.; Bigonciari, C.; Bulgarelli, A.; Camarri, P.; Cardillo, M.; Di Sciascio, G.; Valentina, F.; Marco, R.; Giovanni, P.; Rinaldo, S.; et al. STACEX: A RPC-based detector for a multi-messenger Southern observatory in the GeV-PeV range. In Proceedings of the 37th International Cosmic Ray Conference (ICRC2021), Berlin, Germany, 12–23 July 2021.
79. Acero, F.; Acharya, B.S.; Acín Portella, V.; Adams, C.; Agudo, I.; Aharonian, F.; Al Samarai, I.; Alberdi, A.; Alcubierre, M.; Alfaro, R.; et al. Cherenkov Telescope Array. In Proceedings of the 35th International Cosmic Ray Conference (ICRC2017), Busan, Korea, 10–20 July 2017.

MDPI
St. Alban-Anlage 66
4052 Basel
Switzerland
Tel. +41 61 683 77 34
Fax +41 61 302 89 18
www.mdpi.com

Applied Sciences Editorial Office
E-mail: applsci@mdpi.com
www.mdpi.com/journal/applsci

www.ingramcontent.com/pod-product-compliance
Lightning Source LLC
LaVergne TN
LVHW070601100526
838202LV00012B/530